SpringerBriefs in Electrical and Computer Engineering

T0214228

For further volumes:
http://www.springer.com/series/10059

ChunSheng Xin • Min Song

Spectrum Sharing for Wireless Communications

 Springer

ChunSheng Xin
Electrical and Computer Engineering Department
Old Dominion University
Norfolk, Virginia
USA

Min Song
Computer Science Department
Michigan Technological University
Houghton, Michigan
USA

ISSN 2191-8112   ISSN 2191-8120 (electronic)
SpringerBriefs in Electrical and Computer Engineering
ISBN 978-3-319-13802-2   ISBN 978-3-319-13803-9 (eBook)
DOI 10.1007/978-3-319-13803-9

Library of Congress Control Number: 2015932427

Springer Cham Heidelberg New York Dordrecht London

Printed on acid-free paper

Springer is part of Springer Science+Business Media (www.springer.com)

Acknowledgements

We would like to express our warm appreciation to Old Dominion University and Michigan Technological University. We would like to thank Professor Xuemin "Sherman" Shen and the Springer staff who helped us to publish our work. Special thanks go to U.S. National Science Foundation, in particular the NeTS, EARS, and CAREER programs. Last but certainly not least, we want to thank our families, who never stopped supporting and encouraging us.

Contents

Chapter 1
Introduction

Wireless technology plays a significant role in the national economy. The number of wireless devices has been growing exponentially in recent years, including smart phones, tablets, GPS receivers, baby monitors, remote controllers, wireless sensors, unmanned aircraft systems, etc. Today, there are about 5–7 billion wireless devices. This number is projected to reach 100 billions by 2025, as illustrated in Fig. 1.1. In recent years, the wireless traffic volume has also been doubling about every year. In addition, wireless applications are growing quickly. Besides GPS, cellular phones, WiFi, satellite TV, and public safety communications, new applications are emerging, e.g., smart home, e-health, e-commerce, and intelligent transportation systems, just name a few. The fast growing wireless devices and applications have created an unprecedented demand to radio spectrum. However, the radio spectrum is a limited resource, and has become extremely valuable in recent years, thanks to the ever-increasing demand. Figure 1.2 shows the allocation of radio spectrum in the United States. We can see that almost all of the spectrum that has good propagation property desired by wireless communications has been allocated. Today, there is little spectrum left for future spectrum demands, a problem known as *spectrum scarcity*. In fact, it has become very costly to obtain a license for a new spectrum band. For example, the spectrum auction in 2006 in the United States yielded 13.6 billions for 95 MHz spectrum. The major mobile operator each has spent tens of billion dollars to purchase or trade spectrum in recent years.

The spectrum scarcity problem has a tremendous impact on the national economy, and has drawn the attention from the spectrum regulation agencies, research community, funding agencies, all the way to the highest level. Fortunately, many studies have shown that *the spectrum scarcity is artificial* and mainly created by today's static spectrum allocation policy, rather than the lack of spectrum. As a matter of fact, many studies have found that the licensed spectrum is considerably under-utilized in temporal, spatial, and frequency domains [1, 2]. These findings have spawned a great interest on the research of spectrum sharing, to enable secondary users or devices to access the unused licensed spectrum provided that the primary users of the licensed bands are not harmfully interfered. In the last decade, the FCC, DARPA, NSF, and other agencies have been very active to support related

© The Author(s) 2015   1
C. Xin, M. Song, *Spectrum Sharing for Wireless Communications*,
SpringerBriefs in Electrical and Computer Engineering, DOI 10.1007/978-3-319-13803-9_1

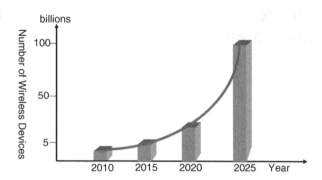

Fig. 1.1 Exponential growth of wireless devices

research including architectures, protocols, algorithms, enabling technologies, radio platforms, and testbeds, through various target programs, including the DARPA XG, WNaN, and SSPARC programs, and the NSF ProWin and EARS programs. Today, spectrum sharing and cognitive communications have become a focus area of major networking and communication conferences, such as IEEE Infocom, ICC, Globecom, to name a few.

1.1 Spectrum Management

Traditionally, the spectrum management takes a static, command and control approach. The spectrum regulation agency usually determines the services to be provided on a spectrum band, e.g., broadcast TV on the band from 470 to 700 MHz.

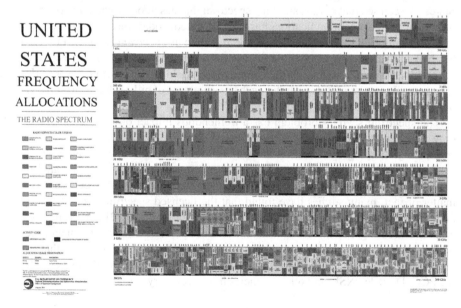

Fig. 1.2 Spectrum allocation map in the United States

The service providers apply for licenses to operate on a spectrum band, to offer the services designated by the spectrum regulation agency. The license application and approval is usually time consuming. Nevertheless, once it is approved, a license can be held for a significant period of time, e.g., 10 or 20 years, and is renewable if needed, to protect the significant investment on the infrastructure by the service provider. The licensed user has exclusive access to the licensed spectrum band.

The static spectrum management has worked well in past decades, and effectively protects the licensed users from interference. However, such static, dedicated, and mutually exclusive spectrum allocation is wasteful, and has created the spectrum scarcity problem. On the other hand, the rapidly proliferated wireless devices and services in recent years have created an ever-increasing demand for radio spectrum. Driven by this fast increasing demand for radio spectrum and the inability of the traditional static spectrum management to address the demand, the spectrum allocation policies have been under reform in recent years, with the objective to allow unlicensed or *secondary users* (SUs) to dynamically access the unused licensed bands, provided that they do not cause harmful interference to the licensed or primary users (PUs). The unused spectrum in the temporal, spatial, and frequency domains are called *spectrum holes* or *white spaces*. They offer a great opportunity for *dynamic spectrum sharing* between SUs and PUs, to relieve the increasing demand for spectrum. Specifically, SUs dynamically search for idle licensed spectrum bands, such as the idle broadcast TV channels (*TV white spaces*), through spectrum sensing or a geo-location database, and access these spectrum bands for data communications [3]. To avoid interference to PUs, SUs need to yield to PUs whenever PUs start using the band.

1.2 Cognitive Radio

The dynamic spectrum sharing between SUs and PUs is made possible by a recent innovation in radio technology–cognitive radio, which is spectrum agile and can sense its spectrum environment and intelligently access idle spectrum bands that are unused by PUs. Traditionally, the radio used for wireless communications is implemented completely by hardware, and typically operates on a fixed spectrum band with a fixed waveform only. In the last decade, the concept of software defined radio or cognitive radio has been proposed in the literature [4–6]. Several cognitive radio platforms have been developed, such as the GNU radio/USRP from the GNU project and the Ettus Research, the WiNC2R radio from WINLAB of the University of New Jersey at Rutgers, the KU Agile Radio from the University of Kansas, and the WARP from Rice University and Mango Communications. The cognitive radio typically consists of a minimal analog RF front-end, analog-to-digital and digital-to-analog converters, and a digital processing engine, which is typically an user-programmable FPGA board, a DSP processor, or even a general-purpose processor, as illustrated in Fig. 1.3. In the case that the digital processing engine is an FPGA or a DSP processor, it is connected to a workstation or a laptop that is used to program the FPGA or the

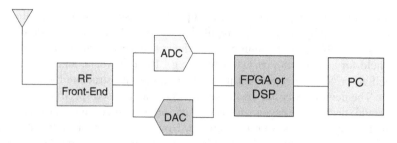

Fig. 1.3 Block diagram of cognitive radio

DSP processor. Most radio functions such as signal processing are implemented in the digital processing engine. Through programming the digital processing engine, the cognitive radio can reconfigure key radio parameters such as the center frequency, bandwidth, waveform, etc., which are traditionally non-configurable in the hardware based radios. As a result, the cognitive radio can dynamically sense spectrum over a wide range of frequencies and search for unused spectrum bands for secondary communications.

1.3 Spectrum Sharing

There have been extensive studies on spectrum sharing in the last decade. These research efforts have showed great promise of dynamic spectrum sharing to significantly increase spectrum utilization. In the following chapters, we will introduce several architectures for spectrum sharing. First of all, we present the original and the de facto standard spectrum sharing architecture, *opportunistic spectrum access*, in Chap. 2. In this architecture, the PU and the SU have mutually exclusive access to a spectrum band. The SU can access the band only when the PU is not using the band. The SU relies on *spectrum sensing* to detect if there is a PU signal on the band. An alternative approach is to utilize a geo-location spectrum database to obtain the information of the available bands; however, this approach is applicable to only certain bands with special features, specifically the TV bands. In general, the spectrum sensing based approach is needed. This is also the approach assumed in most studies on spectrum sharing. The opportunistic spectrum access has a great potential to significantly increase spectrum utilization. Nevertheless, there is a fundamental issue preventing it to realize the full potential of the unused spectrum. The mutually exclusive access to spectrum and the absolute privilege of spectrum access by the PU results in arbitrary disruptions to SU communication by the re-appearance of a PU signal. This results in highly unstable and unpredictable performance and poor quality of service for the SU communications.

To address the issue of arbitrary disruptions to SU communications by the re-appearance of PU signals, we need new architectures for spectrum sharing that

Table 1.1 Comparison of four spectrum sharing architectures

	OSA	IC-DSA	DSCA	ODSA
Methodology	Exclusive access	Network coding	Dirty paper coding	Dynamic allocation
SU transparent to PU?	Yes	No	Yes	No
Disruption to SU from PU?	Yes	No	No	No
Spectrum sharing overhead	High	Low	Low	Low

enable SUs to simultaneously access spectrum with PUs. In Chap. 3, we present an *incentivized cooperative dynamic spectrum access* architecture. In this architecture, *network coding* is conducted between PU traffic and SU traffic and SU nodes serve as relays for PU traffic between PU nodes. While relaying PU packets, SU nodes seek opportunities to encode SU packets onto PU packets for transmission, i.e., SU packets are 'piggybacked' on PU packets without incurring separate spectrum access.

In the incentivized cooperative dynamic spectrum access architecture, the SU communication is not transparent to PUs, as SUs have to use the same waveform as the PUs as they need to decode packets from each other. In Chap. 4, we present a spectrum sharing architecture called *dynamic spectrum co-access*, where the SU communication is *transparent* to the PU. In this architecture, SUs transparently incentivize PUs through increasing the PU performance, by using part of the SU power to increase the *signal to interference plus noise ratio* (SINR) at the PU receiver. As an exchange, SUs can access spectrum simultaneously with PUs since the PU performance is not degraded. Hence there is no disruption to SU communications due to the re-appearance of PU signals.

In Chap. 5, we present an application-oriented spectrum sharing architecture called *on-demand spectrum access*, for users to efficiently share spectrum. In this architecture, a spectrum service provider offers *spectrum services* to users, so that users can dynamically set up *application-oriented* virtual topologies to support user applications. For example, a user can request to set up a virtual topology among a set of nodes to transport a bulk data flow, or carry out a video conference. At last, in Table 1.1, we list a qualitative comparison of the four spectrum sharing architectures: opportunistic spectrum access (OSA), incentivized cooperative dynamic spectrum access (IC-DSA), dynamic spectrum co-access (DSCA), and on-demand spectrum access (ODSA).

References

1. FCC (2003) ET Docket No. 03-222, Notice of Proposed Rule Making and Order, Dec. 2003
2. FCC Spectrum Policy Task Force (2002) ET Docket No. 02-135, Nov. 2002
3. FCC (2004) Unlicensed operation in the TV broadcast bands. ET Docket No. 04-186, Notice of Proposed Rulemaking (NPRM), May 2004

4. Mitola J (1993) Software radios: survey, critical evaluation and future directions. IEEE Aerosp Electron Syst Mag 8(4):25–31
5. Haykin S (2005) Cognitive radio: brain-empowered wireless communications. IEEE J Sel Areas Commun 23(2):201–220
6. Zhao Q, Geirhofer S, Tong L, Sadler B (2008) Opportunistic spectrum access via periodic channel sensing. IEEE Trans Signal Process 56(2):785–796

Chapter 2
Opportunistic Spectrum Access

2.1 Sensing Based Opportunistic Spectrum Access

In the literature, the *opportunistic spectrum access* (OSA) architecture is the de facto architecture for spectrum sharing. Most studies on dynamic spectrum access have assumed this model, e.g., see [1–12] and references therein. The motivation for the OSA architecture is based on the observation that the licensed spectrum is underutilized. Specifically, the licensed spectrum is significantly underutilized in the time, space, and frequency domains. In other words, there are many spectrum holes or white space in the temporal, spatial, and frequency domains, as illustrated in Fig. 2.1a. A spectrum hole is a quiet or idle period of a spectrum band at a certain location. It is also often called white space in the literature. The key feature of the OSA architecture is that the SUs dynamically search for such spectrum holes or white spaces, and opportunistically access spectrum. Figure 2.1b illustrates two snapshots of the spectrum usage at time $t1$ and $t2$, respectively, at a certain location. An SU dynamically finds and selects the spectrum band for access. In the OSA architecture, the PU and the SU have mutual exclusive access to the spectrum band. The PU has the absolute privilege for spectrum access. The SU that is accessing a spectrum band must yield to the PU whenever the PU starts to access the band, in order to avoid harmful interference to the PU. That is, SUs are constrained to opportunistically utilize the spectrum holes or white spaces in the temporal, spatial and frequency domain for communications. The SU uses the cognitive radio to sense the surrounding spectrum environment, then selects one idle spectrum band to transmit data packets. In the OSA architecture, there are three fundamental components: spectrum sensing, spectrum access, and spectrum handoff, which work together to search and access spectrum holes.

2.1.1 Spectrum Sensing

Spectrum sensing is a fundamental component for the OSA architecture. First, before an SU starts communications, it needs to sense spectrum to find an idle spectrum

© The Author(s) 2015
C. Xin, M. Song, *Spectrum Sharing for Wireless Communications*,
SpringerBriefs in Electrical and Computer Engineering, DOI 10.1007/978-3-319-13803-9_2

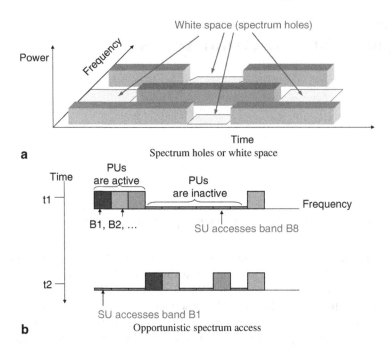

Fig. 2.1 Spectrum holes and the access to spectrum holes by secondary users in opportunistic spectrum access

band to transmit/receive packets. Second, during the SU communications, the SU needs to sense the spectrum band being used for SU communications, to detect if a PU signal re-appears on this band. If there is a PU signal on the band, the SU needs to vacate from the band immediately to avoid harmful interference to PU.

Ideally, spectrum sensing should detect the PU receiver, which is the device we really should protect from SU interference. However, the receiver is often a passive device and hence rather difficult to be detected. Therefore, most studies on spectrum sensing are on the transmitter detection. The spectrum sensing techniques can be approximately classified as in Fig. 2.2. In the next two subsections, we discuss *local sensing* and *cooperative sensing*, respectively.

2.1.1.1 Local Sensing

Local sensing is the spectrum sensing activity performed by each SU individually. There are three major techniques for local spectrum sensing at an SU: energy detection, matched filter detection, and cyclostationary feature detection. The matched filter detection relies on the a priori knowledge of the PU signal, e.g., the modulation, pulse shaping, and the packet format [13]. With such a priori knowledge of the PU signal, an SU can correlate the received signal with a corresponding PU signal, and

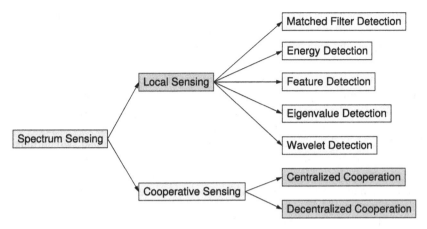

Fig. 2.2 Spectrum sensing techniques

samples the output to detect if the received signal is the PU signal. In contrast, the energy detection does not need any a priori knowledge of the PU signal. It also has other benefits such as low complexity and easy implementation. The received signal energy is measured and compared with a pre-defined threshold [14]. If it exceeds the threshold, then the received signal is treated as a PU signal, and the spectrum band is found busy. Otherwise, the spectrum band is treated as idle. The disadvantage of energy detection is that it is vulnerable to high noise power, which may result in a high false alarm probability. In the scenario with high noise power and/or low signal power, the performance of energy detection may suffer, to result in a high miss-detection probability and/or a high false alarm probability. The cyclostationary feature detection has been developed to address such scenario [15, 16]. This is because, different from noise, most signals have the cyclostationary feature, as their means and autocorrelations exhibit a periodicity feature. Besides these three techniques, there are also other techniques for spectrum sensing such as the eigenvalue detection and the wavelet detection. The eigenvalue detection technique utilizes the eigenvalues of the covariance matrix of the received signal to detect the presence of the PU signal [17]. The wavelet technique is a widely used mathematical tool for analyzing signals. It can also be used to detect the existence of PU signal, particularly in a wide range of spectrum [18].

2.1.1.2 Cooperative Sensing

Local sensing at each individual SU is a starting point for spectrum sensing. Nevertheless, local sensing is often not sufficiently accurate for detecting the PU signal, due to fading, shadowing, and other factors in wireless communications. Therefore, cooperative sensing has been developed to improve the detection accuracy through cooperation among SUs on spectrum sensing. It can operate either in either a decentralized or a centralized mode [19, 20]. With the decentralized mode, SUs exchange local sensing results with each other. After receiving the sensing results from other SUs, an SU can make its own decision on the status of a spectrum band. With the centralized mode, a fusion center such as a base station collects sensing results from all SUs. The fusion center then makes decision on the spectrum band status. The decision making on the spectrum band status can take two approaches: data fusion or decision fusion. With the data fusion approach, all SUs send either raw or processed sensing data, such as the received signal power, to the fusion center, which makes a decision based on the sensing data. With the decision fusion approach, each SU locally processes the sensing data and makes a decision on the band status (often a binary decision of either busy or idle). This decision is then sent to the fusion center. The fusion center makes a final decision based on the local decisions, through a "voting" scheme, e.g., an "OR", "AND", or "Majority" voting.

2.1.2 Spectrum Access and Spectrum Handoff

In the OSA architecture, SUs cannot access a spectrum band when a PU is using it. Moreover, the communication channels (spectrum bands) are dynamically available. Thus, an SU needs to dynamically jump onto different channels over time. That is, an SU usually has to utilize multiple channels for data communication. Therefore, spectrum sharing in the OSA architecture somehow is related to the multi-channel MAC protocols in traditional wireless networks that aim to reduce co-channel interference (and hence increase throughput) by using multiple channels. Nevertheless, the MAC protocols of traditional wireless networks, e.g., the ones in [21, 22], usually cannot be directly used by SUs because they require static channels (i.e., channels are accessible all the time). Hence new protocols and algorithms have been developed for spectrum access for the OSA architecture. These studies primarily take two approaches, depending on whether relying on a control channel. In the first approach (e.g., see [5–12]), a common control channel is used to coordinate SUs for negotiating channels for communications.

The second approach eliminates the dependence on a common control channel among SUs. This is because a common control channel is vulnerable to traffic congestion and jamming attack. However, without a common control channel, the rendezvous between a transmitter and a receiver is a great challenge. For instance, when two SUs are communicating with each other, if the PU signal appears on the channel, the two SUs need to vacate from the channel immediately. In general, they both need to switch to another channel to continue the communication. While the two SUs may negotiate a backup channel before the current channel is disrupted by the PU signal, there is no guarantee that this pre-negotiated backup channel is still available after the current channel becomes unavailable since the availability of every channel is dynamic. The two SUs may frequently check and update the

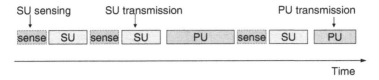

SU sensing   SU transmission   PU transmission

| sense | SU | sense | SU | PU | sense | SU | PU |

Time

Fig. 2.3 Spectrum access in the OSA architecture

backup channel if necessary, to make sure it is up-to-date. However, the overhead for maintaining the backup channel would be high.

A more popular approach is to let an SU hop on channels by a channel hopping sequence, without maintaining a backup channel [23–29]. Instead of continuing the communication with the peer after the current channel becomes unavailable, an SU simply hops to the next channel in the hopping sequence every time slot, and communicates with the SUs currently on that channel. After hopping to the last channel in the sequence, the hopping starts again from the first channel. Such a hopping cycle from the first channel to the last channel is called a *frame*. The communication with the current peer SU is halted, and will be continued when the two SUs hop to the same channel. A critical requirement of this approach is that the channel hopping sequences have to be designed such that any two SUs are ensured to hop to the same channel at some time slot in a frame. Fortunately, this is possible, e.g., through a *quorum* system [25]. Another approach without using a control channel is to estimate the current channel of the receiver SU [30–32]. With this approach, each SU randomly selects the operational channel. The SUs are hence evenly distributed on to all available channels, which minimizes the co-channel interference between SUs, and maximizes throughput. When an SU has packets to another SU, it estimates the current operational channel of the receiver SU, and then switches the radio to the receiver's channel. The channel selection scheme for the operational channel selection by each SU is intelligently designed such that the success probability of channel estimation is high, when an SU estimates the channel of another SU. Hence the transmitter SU can meet the receiver with a high probability.

The spectrum access of SUs generally typically takes a sensing-transmission cyclic mode, as illustrated in Fig. 2.3. After two SUs switch to the same channel, before starting packet transmissions, the SUs first sense the channel to ensure that there is no PU signal. If the channel is idle, the transmitter SU begins to send packets to the receiver for a period of time. Then the SU communication needs to be paused, and the SUs need to sense the channel again to make sure that the PU signal has not become active during this period. Only if the PU signal is not detected, the SU communication resumes. This sensing-transmission cycle is repeated until the SU communication is completed. The periodic spectrum sensing is necessary as when during the SUs packet transmission period, the SUs cannot detect if a PU signal appears on the channel, since the full duplex wireless communication, i.e., sensing while transmitting by the SU, is difficult with today's technology. Whenever a PU signal is detected on the channel, the SUs can wait until the PU signal disappears, or switch to another channel (spectrum handoff) as discussed earlier.

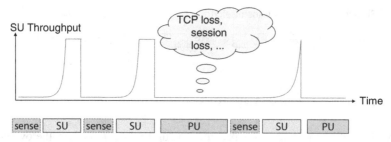

Fig. 2.4 Unstable SU throughput due to disruptions from re-appearance of PU signals

2.1.3   Challenges for the OSA Architecture

In the OSA architecture, SU needs to accurately detect the PU signal in order to avoid harmful interference to the PU. However, by today's technology, accurate spectrum sensing is very challenging, due to multipath, fading, and shadowing, and the ever growing radio interference [33, 34]. Cooperative spectrum sensing can relieve these problems, but also raises new problems such as complicated coordination, increased decision time, and vulnerability to the *spectrum sensing data falsification attack* [35]. Furthermore, spectrum access by SUs in the OSA architecture has to operate with a sensing-transmission cyclic mode, due to the technical difficulty of sensing while transmitting. This introduces significant overhead. Moreover, the re-appearance of PU signal disrupts the SU communications arbitrarily, since the SUs have to vacate from the channel whenever the PU starts using the channel. Such arbitrary disruptions result in highly unstable and unpredictable SU communications, which in turn results in poor quality of service (QoS). Figure 2.4 illustrates a typical SU throughput. Due to the disruptions from the re-appearance of PU signals, and the periodic sensing, the SU throughput fluctuates dramatically, resulting in poor quality of service for applications. The requirement that SUs must yield to the PU whenever the PU starts using the spectrum band also makes SUs vulnerable to the *primary user emulation* attack, where a malicious SU emulates a PU signal by transmitting the PU waveform through a cognitive radio.

2.2   Geo-Location Based Spectrum Access

While the spectrum sensing based OSA architecture is the spectrum sharing architecture assumed by most studies in the literature, for some bands with special usage patterns, an alternate approach based on the geo-location spectrum database can be used. In 2010, FCC officially announced the secondary access to TV white spaces, utilizing a spectrum database approach. Specifically, all the TV broadcasters and wireless microphone users need to register their usage in a spectrum database. For a TV white space device to access spectrum, it first contacts the TV usage database and gets the available channels based on its location.

The frequency of TV white space is seen as "golden standard" frequency for broadcast wireless access service. This is because the TV bands have excellent propagation property and building penetration capability. By the physical law, the radio coverage of a device such as a base station is proportional to the square of the frequency. This translates to that the coverage at TV bands is about 10 times larger than the 1800 MHz cellular band, and 20 time larger than the 2.6 GHz band, which was also a candidate for 4G cellular service. This in turn means it costs significantly less to build a network since the number of base stations at the TV bands are 10 times or 20 times less.

TV bands are below 1 GHz and have excellent properties for propagation, building and foliage penetration, and non-line of sight connectivity. They offer excellent opportunities to support various applications, including wireless broadband access, WiFi-like networks with better coverage and penetration, and traffic offload for another subscription based network such as 4G cellular network.

2.2.1 TV Band Usage

Definition 2.1 A *TV white space channel* is an unused or unoccupied TV channel, i.e., there is no active TV broadcasting on the channel. A *first adjacent (white space) channel* is a white space channel that is right next to an occupied TV channel. A *second adjacent (white space) channel* is a white space channel that is not neighboring to any occupied channel.

In 2008, FCC set rules to allow unlicensed devices to use TV white space. These devices are called *TV white space devices*. The devices are classified into fixed and portable, depending on the transmit power. The fixed devices are allowed to have 4 watts EIRP, while the portable devices are allowed to have either 40 or 100 milliwatts (mW) EIRP, depending on the distance between the operating channel and the closest occupied TV channel which has active TV broadcasting. The fixed devices such as base stations are allowed to operate only in the second adjacent white space channels as defined in Definition 2.1. The portable devices can be allowed to operate in both first and second adjacent channels, with distinction on transmit power. In the first adjacent channels, the transmit power of the portable devices is limited to 40 mW, while in the second adjacent channels, the transmit power of the portable devices is limited to 100 mW. Figure 2.5 illustrates the classification of channels and the allowed operation of the TV white space devices.

2.2.2 TV White Space Availability

Originally, there were totally 83 TV channels. With the channel relocations/repurposing over the years, after the analog to digital TV transition in 2008, there remains only 50 broadcast TV channels, from channel 2 to channel 51. These

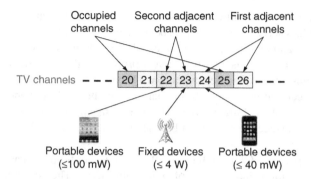

Fig. 2.5 Classification of TV white space channels, and the allowed operation of TV white space devices. The fixed devices such as base stations are allowed to operate on *second adjacent white space channels* only with up to 4 W transmit power. The portable devices can operate in both *first and second adjacent channels*, with the transmit power up to 40 and 100 mW, respectively

channels are distributed on three non-contiguous bands, 54–88 MHz, 174–216 MHz, 470–698 MHz. The fixed TV white space devices are allowed to operate on any TV channel from 2 to 51 as long as it is a second adjacent channel. The portable devices are allowed to operate from channel 21 to 51 only.

Many studies have shown that a large number of TV channels are not used in a vast portion across the United States. However, the use of available TV channels is constrained depending on the device type, as governed by the FCC rules discussed earlier. For instance, the fixed devices, which are essential for commercial applications such as wide area wireless broadband access, cannot use every available TV channel. Based on the rules for the fixed device, the TV channel availability shrinks. One profound observation is that the TV white space available for fixed devices is mostly available in sparsely populated areas, such as the mid-west, and rural areas, while the densely populated metropolitan areas have few available TV channels or even not at all.

2.2.3 TV White Space Access

The current framework of using TV white space is primarily a static approach for spectrum sharing. A key component is to build geo-location spectrum databases to track and assign TV white space channels. The PUs, TV broadcasters, TV translators, and wireless microphones, are all required to register their locations, TV channels under use, and the time periods of active TV channel usage. FCC requires that all spectrum databases have to synchronize with each other so that a PU device needs to register with one database only. An SU that wants to use TV white space has to first register with a geo-location database, by sending its ID, current location, as well as other parameters, through a non-TV white space communication channel across the Internet, e.g., a WiFi connection to the Internet, or a mobile broadband connection such as LTE. As discussed earlier, here an SU refers to either a portable device such as a mobile terminal, or a fixed device such as a base station. After an SU registers successfully with a geo-location database, the database assigns a list of available TV channels that can be used by the device based on the device location, subject to the device type and transmit power, to protect the PUs from harmful interference. The set of channels that can be used by a device may need to be dynamically updated due to mobility of the device, or the change of the TV channel availability. For example, the database may receive a channel registration from a new wireless microphone user.

To facilitate this procedure for some devices that do not have any kind of Internet connection, the devices are classified into *master devices* and *slave devices*. A master device has a non-TV white space Internet connection. Typically, this is a base station with a wired Internet connection. A slave device registers itself with a geo-location database through a master device, and hence does not need to have an Internet connection. Certainly the slave device still needs a non-TV white space connection to a master device. Typically, a slave device is a portable device that has a WiFi connection with a master device which is typically a base station with WiFi capability.

References

1. Zhao Q, Geirhofer S, Tong L, Sadler B (2008) Opportunistic spectrum access via periodic channel sensing. IEEE Trans Signal Process 56(2):785–796
2. Huang S, Liu X, Ding Z (2008) Opportunistic spectrum access in cognitive radio networks. Proceedings of IEEE Infocom, pp 1427–1435
3. Wang B, Ji Z, Liu K (2007) Primary-prioritized Markov approach for dynamic spectrum access. Proc. IEEE DySPAN, pp 507–515
4. Shi Y, Hou T (2008) A distributed optimization algorithm for multi-hop cognitive radio networks. Proceedings of IEEE Infocom, pp 1292–1300
5. Zhao Q, Tong L, Swami A, Chen Y (2007) Decentralized cognitive MAC for opportunistic spectrum access in ad hoc networks: a POMDP framework. IEEE J Sel Areas Commun 25(3):589–600
6. Hamdaoui B, Shin K (2008) OS-MAC: an efficient MAC protocol for spectrum-agile wireless networks. IEEE Trans Mob Comput 7(8):915–930
7. Jia J, Zhang Q, Shen X (2008) HC-MAC: a hardware-constrained cognitive MAC for efficient spectrum management. IEEE J Sel Areas Commun 26(1):106–117
8. Timmers M, Dejonghe A, van der Perre L, Catthoor F (2007) A distributed multichannel MAC protocol for cognitive radio networks with primary user recognition. Proc. Crowncom, pp 216–223
9. Le L, Hossain E (2008) OSA-MAC: a MAC protocol for opportunistic spectrum access in cognitive radio networks. Proc. IEEE WCNC, pp 1426–1430
10. Su H, Zhang X (2008) Cross-layer based opportunistic MAC protocols for qos provisionings over cognitive radio wireless networks. IEEE J Sel Areas Commun 26(1):118–129
11. Yuan Y, Bahl P, Chandra R, Moscibroda T, Wu Y (2007) Allocating dynamic time-spectrum blocks in cognitive radio networks. Proc. ACM MobiHoc, pp 130–139
12. Yuan Y, Bahl P, Chandra R, Chou P, Ferrell J, Moscibroda T, Narlanka S, Wu Y (2007) KNOWS: cognitive radio networks over white spaces. Proc. IEEE DySPAN, pp 416–427
13. Sahai A, Hoven N, Tandra R (2004) Some fundamental limits on cognitive radio. Proc. Allerton Conference

14. Digham F, Alouini M, Simon M (2003) On the energy detection of unknown signals over fading channels. Proc. IEEE ICC
15. Sutton P, Nolan K, Doyle L (2008) Cyclostationary signatures in practical cognitive radio applications. IEEE J Sel Areas Commun 26(1):13–24
16. Turunen V, Kosunen M, Huttunen A, Kallioinen S, Ikonen P, Parssinen A, Ryynanen J (2009) Implementation of Cyclostationary Feature Detector for Cognitive Radios. CROWMCOM'09
17. Zeng Y, Liang Y-C (2009) Eigenvalue-based spectrum sensing algorithms for cognitive radio. IEEE Trans Commun 57(6):1784–1793
18. Tian Z, Giannakis G (2006) A wavelet approach to wideband spectrum sensing for cognitive radios. Cognitive Radio Oriented Wireless Networks and Communications, 2006. 1st International Conference on, pp 1–5
19. Zhao Y, Min S, Xin C (2011) A weighted cooperative spectrum sensing framework for infrastructure-based cognitive radio networks. Comput Commun (Elsevier Computer Science (in press))
20. Gandetto M, Regazzoni C (2007) Spectrum sensing: a distributed approach for cognitive terminals. IEEE J Sel Areas Commun 25(3):546–557
21. Bahl P, Chandra R, Dunagan J (2004) SSCH: slotted seeded channel hopping for capacity improvement in IEEE 802.11 ad-hoc wireless networks. ACM MobiCom
22. So J, Vaidya NH (2004) Multi-channel MAC for ad hoc networks: handling multi-channel hidden terminals using a single transceiver. Proc. ACM MobiHoc, pp 222–233
23. Zhang Y, Yu G, Li Q, Wang H, Zhu X, Wang B (2014) Channel-hopping-based communication rendezvous in cognitive radio networks. IEEE/ACM Trans Netw 22(3):889–902
24. Zhang Y, Li Q, Yu G, Wang B (2011) ETCH: efficient channel hopping for communication rendezvous in dynamic spectrum access networks. Proceedings of IEEE Infocom
25. Bian K, Park J-M (2011) Asynchronous channel hopping for establishing rendezvous in cognitive radio networks. Proceedings of IEEE Infocom, pp 236–240
26. Lin Z, Liu H, Chu X, Leung Y-W (2011) Jump-stay based channel-hopping algorithm with guaranteed rendezvous for cognitive radio networks. Proceedings of IEEE Infocom, pp 2444–2452
27. Shih C-F, Wu TY, Liao W (2010) DH-MAC: a dynamic channel hopping mac protocol for cognitive radio networks. Proc. IEEE ICC
28. Bian K, Park J-M, Chen R (2009) A quorum-based framework for establishing control channels in dynamic spectrum access networks. Proc. ACM MobiCom, pp 25–36
29. Theis N, Thomas R, DaSilva L (2011) Rendezvous for cognitive radios. IEEE Trans Mob Comput 10(2):216–227
30. Xin C, Song M, Ma L, Shetty S, Shen C-C (2010) Control-free dynamic spectrum access for cognitive radio networks. Proc. IEEE ICC
31. Xin C, Song M, Ma L, Shen C-C (2011) Performance analysis of a control-free dynamic spectrum access scheme. IEEE Trans Wirel Commun 10(12):4316–4323
32. Xin C, Song M, Ma L, Shen C-C (2013) Rop: near-optimal rendezvous for dynamic spectrum access networks. IEEE Trans Veh Technol 62(7):3383–3391
33. FCC (2004) Unlicensed operation in the TV broadcast bands. ET Docket No. 04-186, Notice of Proposed Rulemaking (NPRM), May 2004
34. Wellens M, Riihijarvi J, Gordziel M, Mahonen P (2008) Evaluation of cooperative spectrum sensing based on large scale measurements. Proc. IEEE DySPAN
35. Chen C, Song M, Xin C (2013) A density based scheme to countermeasure spectrum sensing data falsification attacks in cognitive radio networks. Proc. GLOBECOME, pp 623–628

Chapter 3
Incentivized Cooperative Dynamic Spectrum Access

3.1 Introduction

More or less, in the OSA architecture, there is a 'foe' relationship between the PU and the SU. Specifically, an SU may access a spectrum band only when the licensed PU is not using it. Moreover, the re-appearance of PU signal disrupts SU communications. As a matter of fact, the OSA architecture primarily benefits SUs and do not offer incentives to PUs to cooperate in dynamic spectrum access. On the other hand, to comply with stringent spectrum access policies, SUs have to behave conservatively to sense and access the spectrum bands, resulting in an actual spectrum utilization significantly lower than what is expected.

We present a new spectrum sharing architecture, called *incentivized cooperative dynamic spectrum access* (IC-DSA), to offer sufficient incentives to PUs so that they cooperate in dynamic spectrum access. This is achieved by utilizing network coding between PU traffic and SU traffic and letting SU nodes serve as relays between PU nodes that are not connected otherwise, or connected with poor quality (e.g., high loss rate) links. Specifically, in the IC-DSA architecture, while relaying PU traffic, SUs also try to encode SU traffic onto PU traffic for transmissions, i.e., SU traffic is 'piggybacked' on PU traffic via network coding, without separate spectrum access[1]. Thus, in the IC-DSA architecture, SUs and PUs create a unique 'win-win' situation for both through network coding and SU relay of PU traffic. The IC-DSA architecture offers increased throughput and reduced packet delay to PUs.

Compared with the OSA architecture, the IC-DSA architecture has the following merits:

Merit 1: PU throughput and spectrum access are guaranteed not degraded, compared with the case of no SU. In fact, the PU can benefit with increased throughput and reduced delay.

---

[1] Note that network coding does not increase the packet size nor the transmission time for the coded packet.

© The Author(s) 2015
C. Xin, M. Song, *Spectrum Sharing for Wireless Communications*,
SpringerBriefs in Electrical and Computer Engineering, DOI 10.1007/978-3-319-13803-9_3

Merit 2: SUs can access spectrum to forward SU traffic even when the PU is active, through coding SU traffic onto PU traffic. In other words, SUs can transmit SU traffic when the PU is either active or inactive. In contrast, in the OSA architecture, SUs can transmit traffic only when the PU is not active.

Merit 3: The performance of both the PU and the SU is improved, compared with the OSA architecture where the PU and the SU compete for exclusive access to spectrum. We call this as a *win-win* situation.

In addition to the above merits, the stringent requirements for spectrum sensing by SUs can also be relieved in the IC-DSA architecture, as PUs can be more tolerable to SU interference, since SUs help to relay part of PU traffic, and thus relieves the PU's burden in traffic forwarding.

3.2 The IC-DSA Architecture

In the IC-DSA architecture, when an SU node switches to a spectrum band, the SU node reconfigures the cognitive radio with the same waveform of the PU in that band. The commonly used PU waveforms are stored in the cognitive radio. When the SU node switches to the spectrum band, it analyzes the waveform used by PU in this band, extracts the features, and determines the PU waveform. Then the cognitive radio switches to this PU waveform. Alternatively, the mapping between spectrum bands and PU waveforms can be stored in SU nodes in advance. An SU node switching to a spectrum band searches the mapping table and configures the cognitive radio with the corresponding waveform. In the IC-DSA architecture, SUs and PUs operate as follows:

- Each SU node listens to both PU and SU packets and relay them. PU packets have a higher priority to be relayed.
- When relaying a PU packet, an SU node not only tries to use network coding to encode other PU packet with this packet, but also tries to use network coding to 'piggyback' one or more SU packets onto this PU packet.
- When there is no PU packet in its queue and hence there is no piggybacking opportunity for SU packets, an SU node opportunistically transmits SU packets whenever possible, as in the OSA architecture.
- When an SU node tries to send either a PU packet or a PU packet encoded with SU packets, the SU node has the same spectrum access right as PU nodes. When an SU node tries to send only SU packets, the SU node has to wait until there is no PU activity, as the OSA architecture.
- In addition to the packets from PU nodes, each PU node also listens to packets from SU nodes. If a received packet is a coded packet, then the PU node decodes it to get PU packets intended to it. If it is an SU packet, then the PU node stores it as it might be needed for future decoding.

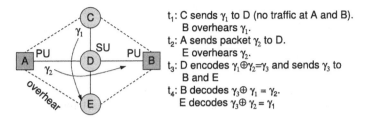

t_1: C sends γ_1 to D (no traffic at A and B). B overhears γ_1.

t_2: A sends packet γ_2 to D. E overhears γ_2.

t_3: D encodes $\gamma_1 \oplus \gamma_2 = \gamma_3$ and sends γ_3 to B and E

t_4: B decodes $\gamma_3 \oplus \gamma_1 = \gamma_2$. E decodes $\gamma_3 \oplus \gamma_2 = \gamma_1$

Fig. 3.1 Illustration of the IC-DSA architecture

IC-DSA ensures that both the PU and the SU can decode packet(s) intended to themselves, respectively. Figure 3.1 illustrates the IC-DSA architecture. SU node D serves as a relay for PU packets and opportunistically transmits SU packet without degrading PU throughput. In the figure, node A may or may not reach node B. In the former case, due to the lossy nature of wireless links, the packet from A can reach B with a delivery probability q. Node D can still relay the packet to B, to increase the total delivery probability.

3.3 Analysis of IC-DSA Performance

We aim to find the maximum achievable PU and SU performance in IC-DSA. We use throughput as the utility function. Nevertheless, the model can be extended to other utility functions as well. We adopt the *throughput* definition used in several optimization works (e.g., see [1–3]). Specifically, let the traffic demand of node pair k ($1 \leq k \leq K$) be α_k Mbps, the throughput is defined as a scaling factor $\lambda \geq 0$ such that, when changing the traffic load of node pair k from α_k to $\lambda\alpha_k$, for all $1 \leq k \leq K$, the network can support these traffic loads. We refer to $\lambda\alpha_k$ as the actual traffic load of node pair k when the network throughput is λ.

We consider unicast demands and use inter-session network coding. Let $\mathcal{D}$ and $\bar{\mathcal{D}}$ denote the set of PU and SU node pairs, respectively. Let $\mathcal{P}_k$ denote the set of routing paths for node pair $k \in \mathcal{D} \cup \bar{\mathcal{D}}$. For $k \in \bar{\mathcal{D}}$, the paths in $\mathcal{P}_k$ consist of SU nodes only, since we assume PU nodes do not forward SU traffic. Furthermore, in the OSA architecture, the paths in $\mathcal{P}_k$ for $k \in \mathcal{D}$ consist of PU nodes only, since PUs and SUs operate separately. Let $\mathcal{P} = \bigcup_{k \in \mathcal{D}} \mathcal{P}_k$, and $\bar{\mathcal{P}} = \bigcup_{k \in \bar{\mathcal{D}}} \mathcal{P}_k$. (Note that $\mathcal{P} \cap \bar{\mathcal{P}} = \varnothing$.) The traffic of node pair k is carried on paths in $\mathcal{P}_k$. We call the traffic carried on path $p \in \mathcal{P}_k$ as the flow on path p, denoted as $f(p)$. Let $\mathcal{V}$ denote the PU node set; $\tilde{\mathcal{V}}$ denote the SU node set such that each $v \in \tilde{\mathcal{V}}$ is on some PU routing path $p \in \mathcal{P}$; and $\tilde{\mathcal{V}}'$ denote the SU node set such that for each $v \in \tilde{\mathcal{V}}'$, all routing paths going through node v are in $\bar{\mathcal{P}}$, i.e., only SU traffic goes through v.

The major factor to determine throughput is the amount of traffic each node can transmit. In wireless networks, packet transmission at one node may interfere other nodes. We assume the *protocol interference model* [4]. Let $\mathcal{N}_v$ denote the neighbor nodes of node v. The broadcast nature of wireless communications can be modeled by representing the packet transmission as a hyperarc (v, J), with node v as the sender and a subset of neighbors $J \subseteq \mathcal{N}_v$ as receivers. Let $\mathcal{H}_v = \{(v, J) \mid J \subseteq \mathcal{N}_v\}$ denote the set of hyperarcs of node v. Let $\mathcal{I}_v$ denote the set of nodes that node v interferes. The packet transmission can be modeled as a *conflict graph*, where each vertex is a hyperarc $(v, J) \in \cup_{v \in \mathcal{V} \cup \bar{\mathcal{V}} \cup \bar{\mathcal{V}}'} \mathcal{H}_v$, and there is an edge between two vertices, (v, J) and $(i, \mathcal{S})$, if node v interferes some node in $\mathcal{S}$ or node i interferes some node in J. There is a fundamental *interference constraint* for packet transmissions in wireless networks: among a set of hyperarcs that interfere each other, at most one of them can transmit at a given time. This constraint is equivalent to the following: the summation of the fraction of transmission time of all hyperarcs in a clique cannot be larger than 1. The fraction of transmission time by a hyperarc is computed as the hyperarc traffic load divided by the data rate of the hyperarc, which is approximately the channel capacity if wireless links are in good quality. In the following, we will derive the traffic load of each hyperarc, using *linear programing* (LP).

We introduce the concept of *coding mode*. Our formulation can use both a COPE-like network coding approach [5, 6], and a new network coding approach to be described in next section. In network coding, each packet transmitted by a node may be a *native* or uncoded packet, or a *coded* packet that encodes several *component* packets. The coding mode of a coded packet transmission is characterized by the incoming and outgoing links of the component packets. For example, let $\gamma_1 \oplus \cdots \oplus \gamma_h$ ($h \geq 2$) denote a packet encoded by a node. Suppose component packet γ_j ($1 \leq j \leq h$) comes from a neighbor connected by link e_j^{in} and goes to a neighbor connected by link e_j^{out}. A link-pair $\ell_j = (e_j^{in}, e_j^{out})$ is used to indicate this incoming-outgoing relationship. A parameter $a_j = $ 'n' or 'c' indicates whether packet γ_j is received by this node as a native packet or inside a coded packet. The coding mode of the coded packet $\gamma_1 \oplus \cdots \oplus \gamma_h$ is denoted as $[(\ell_1, a_1), \ldots, (\ell_h, a_h)]$.

Let $\mathcal{M}_v$ denote the set of all possible coding modes at node $v \in \mathcal{V} \cup \bar{\mathcal{V}}'$, i.e., when node v has either PU traffic or SU traffic, but not both. Let $\mathcal{U}_v$ denote the set of coding modes for $v \in \bar{\mathcal{V}}$, i.e., when node v has both PU traffic and SU traffic. For each coding mode $m \in \mathcal{U}_v$, we add an extra element to each link-pair of m, $b_j = $ 'p' or 's', to indicate if the traffic transported on this link-pair is PU or SU traffic. In other words, we transform each coding mode $m = [(\ell_1, a_1), \ldots, (\ell_h, a_h)]$ into a set of new coding modes $\{[(\ell_1, a_1, b_1), \ldots, (\ell_h, a_h, b_h)]\}$, where $b_j = $ 'p' if only PU routing paths go through ℓ_j; $b_j = $'s' if only SU routing paths go through ℓ_j; and $b_j \in \{$'p','s'$\}$ otherwise. For example, suppose we have $m = [(\ell_1, a_1), (\ell_2, a_2)]$. Only PU routing paths go through ℓ_1, and both PU and SU routing paths go through ℓ_2. Then coding mode m is transformed into two modes, $[(\ell_1, a_1, \mathsf{p}), (\ell_2, a_2, \mathsf{p})]$, and $[(\ell_1, a_1, \mathsf{p}), (\ell_2, a_2, \mathsf{s})]$. For the ease of description, we use notations $s(\cdot)$ and $d(\cdot)$ to denote the source and destination, respectively, of a path, a link-pair, or a coding mode. The source (destination) nodes of a coding mode are the set of source (destination) nodes of all link-pairs, i.e., $s(m) = \{s(\ell_j) \mid \ell_j \in m\}$ and $d(m) = \{d(\ell_j) \mid \ell_j \in m\}$. $\mathcal{M}_v$ and $\mathcal{U}_v$ can be obtained through exhaustive search of all subsets of node v's link-pairs that are traversed by routing paths. Specifically, a subset of link-pairs $[\ell_1, \ldots, \ell_h]$ is a coding mode if for any i $(1 \le i \le h)$, $d(\ell_i)$ overhears $s(\ell_j)$ for all $j \ne i, 1 \le j \le h$. This is implied by the *coding application condition* discussed in the next section. The component a_i for each link-pair in a coding mode is equal to 'n' only, except in the case that $h = 2$, and $s(\ell_1) = d(\ell_2)$, where $a_1 = \{\text{'n'},\text{'c'}\}$; or $s(\ell_2) = d(\ell_1)$, where $a_2 = \{\text{'n'},\text{'c'}\}$; or both are true, where $a_1 = a_2 = \{\text{'n'},\text{'c'}\}$.

In the following, we list notations and then present an LP model to obtain throughput.

$\mathcal{P}_k, \mathcal{P}, \bar{\mathcal{P}}$	Paths defined above.
$\mathcal{V}, \bar{\mathcal{V}}, \bar{\mathcal{V}}', \mathcal{N}_v, \mathcal{H}_v, \mathcal{M}_v, \mathcal{U}_v$	Nodes, neighbors, hyperarcs, and coding modes, defined above.
$\mathcal{C}$	Set of all maximal cliques in the hyperarc conflict graph.
$\mu_v(J)$	Transmission rate of hyperarc (v, J).
$\mathcal{L}_v$	Set of link-pairs of node v. Each wireless link incident to node v is seen as one incoming and one outgoing link.
α_k	Demand of node pair $k \in D$.
β_k	Demand of node pair $k \in \bar{D}$.
η	PU throughput.
λ	SU throughput.
$f(p)$	Flow on path p, i.e., the end-to-end traffic load carried on path p.
$x(m)$	Traffic load transported by coding mode $m \in \mathcal{M}_v$.
$u(m)$	Traffic load transported by coding mode $m \in \mathcal{U}_v$.
$z_v(p)$	Traffic load of flow $f(p)$ that is sent by node v as native packets (see Fig. 3.3).
$y_v(J)$	Traffic load sent by node v to hyperarc (v, J). This is the amount of traffic sent to the set J exactly, excluding the traffic sent to a hyperarc that is a superset of (v, J).

Figure 3.2 illustrates an LP model, which can be used to maximize either PU or SU throughput, with the other as a free variable, or as a constraint. In the LP model, the upper bound for all variables is ∞ except variable $y_v(J)$, which is bounded by $\mu_v(J)$. Constraints (3.1)—(3.13) together derive the load on each hyperarc, $y_v(J)$. The throughput is maximized subject to the interference constraint (3.14). Constraint (3.1) is flow conservation on routing paths. Constraint (3.2) is flow conservation on each link-pair of node $v \in \mathcal{V} \cup \bar{\mathcal{V}}'$, as shown in Fig. 3.3. The left side is the total traffic load of all flows transiting node v on link-pair ℓ, whereas in the right side, the first term is the traffic load that is received as either native packets or coded packets by node v, and sent to the downstream node $d(\ell)$ as native packets; the second term $\sum_{m \in \mathcal{M}_v,(\ell,a) \in m} x_v(m)$ is the traffic load that is received as either native ($a = \text{'n'}$) or coded ($a = \text{'c'}$), and sent to $d(\ell)$ as coded by node v. Constraints (3.3)—(3.4) are flow conservation on the PU and SU traffic, respectively, for the link-pair of node $v \in \bar{\mathcal{V}}$, whose coding modes distinguish PU and SU traffic. Constraint (3.5) limits the traffic load of a flow transmitted as native packets by each node on the path. Constraint (3.6) limits the traffic load that is received as native but sent as coded

maximize λ (or η), **subject to**

$$\forall k \in \bar{\mathscr{D}}, \sum_{p \in \mathscr{P}_k} f(p) = \lambda \beta_k, \quad \forall k \in \mathscr{D}, \sum_{p \in \mathscr{P}_k} f(p) = \eta \alpha_k \tag{3.1}$$

$\forall v \in \mathscr{V} \cup \bar{\mathscr{V}}'$ and $\ell \in \mathscr{L}_v$,

$$\sum_{\substack{p \in \mathscr{P} \cup \bar{\mathscr{P}} \\ \ell \in p}} f(p) = \sum_{\substack{p \in \mathscr{P} \cup \bar{\mathscr{P}} \\ \ell \in p}} z_v(p) + \sum_{a \in \{'n','c'\}} \sum_{\substack{m \in \mathscr{M}_v \\ (\ell,a) \in m}} x(m) \tag{3.2}$$

$\forall v \in \bar{\mathscr{V}}$ and $\ell \in \mathscr{L}_v$,

$$\sum_{\substack{p \in \mathscr{P} \\ \ell \in p}} f(p) = \sum_{\substack{p \in \mathscr{P} \\ \ell \in p}} z_v(p) + \sum_{a \in \{'n','c'\}} \sum_{\substack{m \in \mathscr{U}_v \\ (\ell,a,'p') \in m}} u(m) \tag{3.3}$$

$$\sum_{\substack{p \in \bar{\mathscr{P}} \\ \ell \in p}} f(p) = \sum_{\substack{p \in \bar{\mathscr{P}} \\ \ell \in p}} z_v(p) + \sum_{a \in \{'n','c'\}} \sum_{\substack{m \in \mathscr{U}_v \\ (\ell,a,'s') \in m}} u(m) \tag{3.4}$$

$$\forall p \in \mathscr{P} \cup \bar{\mathscr{P}}, z_{s(p)}(p) = f(p), z_v(p) \le f(p) \text{ for } v \in p \setminus \{s(p), d(p)\} \tag{3.5}$$

$\forall v \in \mathscr{V} \cup \bar{\mathscr{V}}'$ and $\ell \in \mathscr{L}_v$,

$$\sum_{m \in \mathscr{M}_v, (\ell,'n') \in m} x(m) \le \sum_{p \in \mathscr{P} \cup \bar{\mathscr{P}}, \ell \in p} z_{s(\ell)}(p) \tag{3.6}$$

$$\sum_{m \in \mathscr{M}_v, (\ell,'c') \in m} x(m) \le \sum_{p \in \mathscr{P} \cup \bar{\mathscr{P}}, \ell \in p} \left(f(p) - z_{s(\ell)}(p) \right) \tag{3.7}$$

$\forall v \in \bar{\mathscr{V}}$ and $\ell \in \mathscr{L}_v$,

$$\sum_{\substack{m \in \mathscr{U}_v, \\ (\ell,'n','p') \in m}} u(m) \le \sum_{\substack{p \in \mathscr{P}, \\ \ell \in p}} z_{s(\ell)}(p), \quad \sum_{\substack{m \in \mathscr{U}_v, \\ (\ell,'n','s') \in m}} u(m) \le \sum_{\substack{p \in \bar{\mathscr{P}}, \\ \ell \in p}} z_{s(\ell)}(p) \tag{3.8}$$

$$\sum_{m \in \mathscr{U}_v, (\ell,'c','p') \in m} u(m) \le \sum_{p \in \mathscr{P}, \ell \in p} \left(f(p) - z_{s(\ell)}(p) \right) \tag{3.9}$$

$$\sum_{m \in \mathscr{U}_v, (\ell,'c','s') \in m} u(m) \le \sum_{p \in \bar{\mathscr{P}}, \ell \in p} \left(f(p) - z_{s(\ell)}(p) \right) \tag{3.10}$$

$\forall v \in \mathscr{V} \cup \bar{\mathscr{V}} \cup \bar{\mathscr{V}}'$ and $(v,J) \in \mathscr{H}_v$ with $|J| = 1$,

$$y_v(J) = \sum_{p \in \mathscr{P} \cup \bar{\mathscr{P}}, (v,J) \in p} z_v(p) \tag{3.11}$$

$\forall v \in \mathscr{V} \cup \bar{\mathscr{V}}'$ and $(v,J) \in \mathscr{H}_v$ with $|J| > 1$,

$$y_v(J) = \sum_{m \in \mathscr{M}_v, J = d(m)} x(m) \tag{3.12}$$

$$\forall v \in \bar{\mathscr{V}} \text{ and } (v,J) \in \mathscr{H}_v \text{ with } |J| > 1, y_v(J) = \sum_{m \in \mathscr{U}_v, J = d(m)} u(m) \tag{3.13}$$

$$\forall c \in \mathscr{C}, \sum_{(v,J) \in c} y_v(J) / \mu_v(J) \le 1 \tag{3.14}$$

Fig. 3.2 Maximize PU or SU throughput

by node $v \in \mathcal{V} \cup \bar{\mathcal{V}}'$, which cannot be larger than the traffic load sent as native by $s(\ell)$ (see Fig. 3.3). Similarly, constraint 3.7) limits the traffic load that is both received and sent as coded by node v. Constraint (3.8) limits PU and SU traffic loads, respectively, which are received as native but sent as coded by node $v \in \bar{\mathcal{V}}$. Constraints (3.9)–(3.10) limit PU and SU traffic loads, respectively, which are both

Fig. 3.3 Traffic amount received & sent by node v on link-pair ℓ

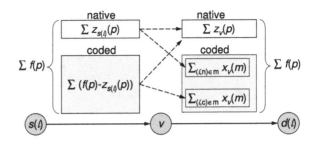

received and sent as coded by node $v \in \bar{\mathcal{V}}$. Constraint (3.11) calculates the natively transmitted PU and SU traffic loads intended to a single neighbor. Constraint (3.12) computes the coded PU and SU traffic loads broadcast to multiple neighbors by a PU node, or an SU node not on PU paths. Constraint(3.13) calculates the same quantity broadcast by an SU node on some PU path. Constraint (3.14) represents the interference constraint in wireless communication, where $\frac{y_v(J)}{\mu_v(J)}$ is the fraction of time that hyperarc (v, J) is transmitting.

3.4 A New Network Coding Scheme

Most studies on network coding for wireless networks (e.g., [3, 5–7]) consider encoding native packets only. We call such a scheme as *coding over native packets* (CoN). Specifically, when a node receives a coded packet, it first decodes the coded packet to extract the expected native packet, and then seeks to encode it with other stored native packets. The advantage is that, while conducting network coding, a node can easily check if each receiver can decode to get the intended packet. For example, in COPE [5, 6], a node periodically notifies the neighbors about its stored packets, via a *packet report*. When a node wants to encode n native packets, $\gamma_1, \ldots, \gamma_n$, to n receivers, it simply check if each receiver already has stored all packets $\gamma_1, \ldots, \gamma_n$ except the one intended to it, e.g., the receiver expecting γ_1 must have stored packets $\gamma_2, \ldots, \gamma_n$ [2]. This is called the *coding application condition* by [7], and the packets stored in a node are referred to as *key set*. However, although it is easy to check the application condition, CoN misses many coding opportunities. This is because for a node, say node A, to encode packets $\gamma_1, \ldots, \gamma_n$ into a coded packet Γ, packets $\gamma_1, \ldots, \gamma_n$ have to be received as native by node A (except the special case of two link-pairs in the exactly opposite direction). Otherwise, the receivers would not be able to decode Γ. For example, Fig. 3.4 shows two bidirectional flows intersecting at node C. Node D has two packets, γ_1, γ_2. Packet γ_1 goes toward node B, and γ_2 has come from B and goes toward the reverse direction. Node D broadcasts the coded

---

[2] This implies that in a coded packet, a node can encode at most one (native) packet intended to a receiver.

Fig. 3.4 Node C cannot
encode two packets, $\gamma_1 \oplus \gamma_2$
and $\gamma_3 \oplus \gamma_4$, which have
arrived as 'coded'

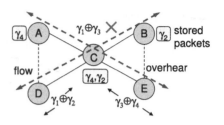

packet $\gamma_1 \oplus \gamma_2$. This coded packet is overheard by node A and stored as a key. Similarly, node E broadcasts the coded packet $\gamma_3 \oplus \gamma_4$. Node C decodes $\gamma_1 \oplus \gamma_2$ to get γ_1, and decodes $\gamma_3 \oplus \gamma_4$ to get γ_3. However, node C cannot send the coded packet $\gamma_1 \oplus \gamma_3$ to A and B, because the key set of node A does not have γ_1 ($\gamma_1 \oplus \gamma_2$ is a different packet), and thus cannot decode $\gamma_1 \oplus \gamma_3$.

The above example illustrates a disadvantage of CoN: the continual network coding between adjacent nodes is generally not possible. In other words, among two adjacent nodes on a path, e.g., nodes D and C in Fig. 3.4, only one of them can apply network coding to the packets on the path. Hence, CoN may miss many coding opportunities.

We present a new network coding approach, called *Coding over Coded Packet* (CoC), to utilize new coding opportunities. We let a node not only encode native packets, but also encode the received coded packets directly (i.e., without 'breaking' them into component packets). For example, in Fig. 3.4, if node C encodes directly over the received packets as $(\gamma_1 \oplus \gamma_2) \oplus (\gamma_3 \oplus \gamma_4)$, then node A can decode this packet to extract γ_3 because it has $(\gamma_1 \oplus \gamma_2)$ and γ_4 in its key set.

We go beyond the intuitive observation in Fig. 3.4 to develop a general solution. We let a node collect all coded and native packets overheard (or received) into its key set. To reduce the size of the key set, a node periodically purges out-of-date packets, and adds a new packet only if it is linearly independent with existing packets in the key set, which can be checked by Gaussian-Jordan elimination (to be discussed later). With CoC, a node can encode an arbitrary composition of native and coded packets. Next we derive the coding application condition for CoC, to ensure that each receiver is able to decode, because the application condition in existing studies [5–7] is not applicable any more. Note that with CoC, although a receiver may just directly encode a received coded packet, we still make sure that the receiver is able to decode the coded packet, so that a native packet transmitted as coded is decodable at each hop, to guarantee the decodability at the destination. We use lowercase γ to denote a native packet and uppercase Γ to denote a coded packet. Suppose node A wants to compose a coded packet $\Gamma_0 = \sum_{j=1}^{n} C(0,j)\gamma_j$, where $\sum$ is the summation on field F_2 (the field with 0 and 1 only), coefficient $C(0,j) = 0$ or 1, and n is the total number of unique native packets that are contained in the keys of node A and its neighbors. We assume that node B is one of the receivers, and the γ_i in Γ_0 is the intended packet to node B. Denote the key set of node B as $\mathcal{K}_B = \{\Gamma_1, \ldots, \Gamma_{|\mathcal{K}_B|}\}$, where $\Gamma_k = \sum_{j=1}^{n} C(k,j)\gamma_j$ $(1 \le k \le |\mathcal{K}_B|)$. The composition of each packet Γ_k in node B's key set has been previously sent from node B to node A, within a packet report. Let M denote the coefficients matrix of packets Γ_0 and $\Gamma_1, \ldots, \Gamma_{|\mathcal{K}_B|}$, with the kth row $(C(k,1), C(k,2), \cdots, C(k,n))$ indicating Γ_k $(0 \le k \le |\mathcal{K}_B|)$. Node A can apply the Gauss-Jordan elimination (on field F_2) to matrix M to get the *reduced echelon row form*, denoted as M', and then knows whether node B can decode Γ_0 to get γ_i by the following theorem.

Theorem 3.1 *Packet Γ_0 can be decoded by node B to extract the intended packet γ_i iff in matrix M', there exists a row that has exactly one non-zero element, and this non-zero element is in column i (corresponding to γ_i).*

Proof The Gauss-Jordan elimination uses three operations: row switching, row multiplication (by non-zero constant), and row addition. The row switching operation exchanges positions of two packets. The row multiplication operation becomes a trivial operation in field F_2, since the only non-zero constant is 1. In field F_2, the row addition (adding a multiplication of one row to another) becomes the XOR operation. Similar to M, the reduced echelon row form M' also represents a coefficients matrix of a list of packets. We denote them as $\{\Upsilon_0, \ldots, \Upsilon_{|\mathcal{K}_B|}\}$. In other words, the Gauss-Jordan elimination transforms a list of packets $\{\Gamma_0, \ldots, \Gamma_{|\mathcal{K}_B|}\}$ into another list of packets $\{\Upsilon_0, \ldots, \Upsilon_{|\mathcal{K}_B|}\}$ through a series of XOR operations, and packets position exchanging. Now if in M', there exists a row $\bar{k}$ which has exactly one non-zero element, and this non-zero element is on column i, then the packet corresponding to row $\bar{k}$, i.e., $\Upsilon_{\bar{k}}$, contains γ_i only, since in row $\bar{k}$, the coefficient for other native packet γ_j $(j \ne i)$ is 0. Thus node B can use its key set to decode Γ_0 and extract γ_i, which is the $\bar{k}$ packet in the list $\{\Upsilon_0, \ldots, \Upsilon_{|\mathcal{K}_B|}\}$ obtained by the Gauss-Jordan elimination. This proves that the condition is sufficient. Note that the series of XOR (row addition) operations during the Gauss-Jordan elimination records the actual decoding procedure, i.e., which packet should be XORed to which packet in each step. Next we prove that the condition is also necessary. Suppose node B can decode Γ_0. This means that starting from some packet in $\{\Gamma_0, \ldots, \Gamma_{|\mathcal{K}_B|}\}$, say Γ_k, we can XOR a sequence of packets from $\{\Gamma_0, \ldots, \Gamma_{|\mathcal{K}_B|}\}$ onto Γ_k to get γ_i, i.e., $\Gamma_k \oplus \Gamma_{t_1} \oplus \cdots \oplus \Gamma_{t_u} = \gamma_i$. We use γ_i to replace Γ_k, to get a new packet list, $\{\Gamma_0, \ldots, \Gamma_{k-1}, \gamma_i, \Gamma_{k+1} \ldots \Gamma_{|\mathcal{K}_B|}\}$. This is fine because Γ_k is linearly dependent with the new packet list, and thus can be safely dropped from the list. Let Q denote the coefficients matrix of

$$\{\Gamma_0, \ldots, \Gamma_{k-1}, \gamma_i, \Gamma_{k+1} \ldots \Gamma_{|\mathcal{K}_B|}\}.$$

It is clear that matrix Q has been transformed from matrix M using row additions, i.e., adding other rows onto row k. Furthermore, since row k represents the coefficient of γ_i, we have $Q(k,i) = 1$ and $Q(k,j) = 0$ for $j \ne i$, i.e., row k has only one non-zero element and it is on column i. Now we obtain the reduced echelon row form of Q. First we use row k to eliminate all 1's in column i of matrix Q, since $Q(k,i) = 1$, and $Q(k,j) = 0$ for $j \ne i$. Now in column i, the only non-zero element is also $Q(k,i) = 1$, with $Q(u,i) = 0$ for $u \ne k$. Next, we take out row k and column i from matrix Q. The remaining matrix is denoted as $\bar{Q}$. We use the Gauss-Jordan elimination to get the reduced echelon row form of $\bar{Q}$, denoted as $\bar{Q}'$. Now we insert the row k and column i that have been taken out from Q back into $\bar{Q}'$, and denote

Fig. 3.5 A scenario where CoC can help while CoN cannot

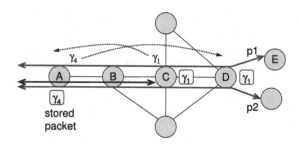

it Q', which will still be in the reduced echelon row form, as column i has only one non-zero element $Q(k, i)$. Since Q has been transformed from M, and the reduced echelon row form is unique, we have $Q' = M'$. Therefore, if node B can decode Γ_0, then M' has a row that has exactly one non-zero element, and this non-zero element is in column i. This finishes the proof.

Remark: The Gauss-Jordan elimination can also be used to check if a list of packets are linearly dependent, which is true iff M' includes a row of all 0's.

Theorem 3.1 can be used in actual network operation to exploit more coding opportunities, to enable continual network coding between adjacent nodes. Nevertheless, it is not easy to directly use it to find coding modes for the LP formulation in the preceding section. Next we derive a corollary that can be used for the LP formulation. Let us examine a scenario illustrated in Fig. 3.5. We consider a coding mode of node C that involves node B as a sender node of a link-pair, B→C→D. We assume that paths are bidirectional. Nevertheless the following discussion is also applicable to unidirectional paths with minor changes. In Fig. 3.5, suppose the routing paths transiting node B all come from one node, A, and either terminate at node C, or go to the destination of the link pair, which is node D in our example. Then the traffic transport at node B is the information exchange scenario [8]. Suppose packet γ_1 comes to node B from node C, and is going to node A. On the other hand, packet γ_4 comes to node B from node A and is going to node D. Packet γ_1 may have been sent by node C as either coded or native. In both cases, node D has γ_1 in its key set[3]. Node B would send the coded packet $\gamma_1 \oplus \gamma_4$ to nodes A and C. Suppose node C encodes the received coded packet $\gamma_1 \oplus \gamma_4$ with other packet(s), e.g., $\gamma_2 \oplus \gamma_3$, which was overheard by node D. By Theorem 3.1, node D can decode $(\gamma_1 \oplus \gamma_4) \oplus (\gamma_2 \oplus \gamma_3)$,

---

[3] If γ_1 has been sent as native by node C, then node D must have overheard it. If γ_1 has been sent as coded, then it should not have been initiated from node C, since a node cannot encode a packet locally initiated from itself. Thus in this case, γ_1 must have come from node D, since all transit paths of node B either terminate at node C or go to node D. Packet γ_1 may come to node D as coded from node E. However, the coding application condition ensures that node D (actually every node on the path) can decode the coded packet to get γ_1. Hence node D has γ_1 in the key set.

Fig. 3.6 The sample topology

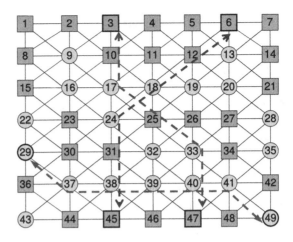

since we have

$$M = \begin{pmatrix} 1 & 1 & 1 & 1 \\ 0 & 1 & 1 & 0 \\ 1 & 0 & 0 & 0 \end{pmatrix} \xrightarrow[\text{Jordan}]{\text{Gauss-}} M' = \begin{pmatrix} 1 & 0 & 0 & 0 \\ 0 & 1 & 1 & 0 \\ 0 & 0 & 0 & 1 \end{pmatrix},$$

where M denotes the coefficients of $\Gamma_0 = (\gamma_1 \oplus \gamma_4) \oplus (\gamma_2 \oplus \gamma_3)$, $\Gamma_1 = \gamma_2 \oplus \gamma_3$, and $\Gamma_2 = \gamma_1$. The 3rd row of M' meets the condition of Theorem 3.1, and thus node D can decode $(\gamma_1 \oplus \gamma_4) \oplus (\gamma_2 \oplus \gamma_3)$ to get γ_4. Hence, node B can send a packet to node C as either coded or native.

Corollary 3.1 *In a coding mode of a node, say node C, if for the source node of a link-pair, all transit paths of this source node come from one node, and either terminate at node C, or go to the destination of the link-pair, then the source node can send packet to node C as either coded or native.*

We use Corollary 3.1 to find CoC coding modes and use them in the LP formulation. The obtained throughput is accordingly called *CoC throughput*.

3.5 Numerical Results

We use CPLEX to solve the LP formulation to obtain throughput. The sample topology is illustrated in Fig. 3.6. We aim to find a *win-win zone* for the PU and the SU in the IC-DSA architecture, i.e., the PU and SU throughputs that are achievable and win-win for both the PU and the SU. The channel bandwidth is assumed 50 Mbps. The hyperarc transmission rate $\mu_v(J)$ is assumed the same as the channel bandwidth. The routing paths are bidirectional. We use both uniform and random demands between node pairs. In the former case, the demand between each node pair is 1 Mbps, while in the latter case, it is a random real number in the range

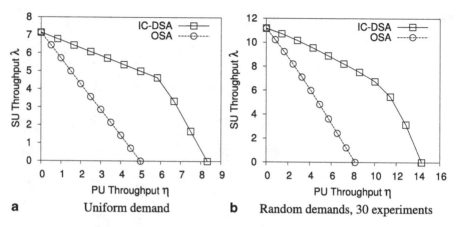

Fig. 3.7 Win-win zone of the PU and the SU

$[0, 1]$. Each node is assumed to interfere with nodes up to 2 hops. We focus on three node pairs, $(3, 47)$, $(6, 45)$, and $(29, 49)$, each with demand 1 (uniform demand). The traffic between other node pairs are treated as background link traffic, which is assumed 1 for each link. The routing paths in the IC-DSA architecture for the three node pairs are shown in Fig. 3.6. In the OSA architecture, the path of node pair $(3, 47)$, denoted as $\langle 3, 47 \rangle$, is $(3, 2, 8, 15, 23, 30, 36, 44, 45, 46, 47)$, the path $\langle 6, 45 \rangle$ is $(6, 14, 21, 27, 34, 42, 48, 47, 46, 45)$, and the path $\langle 29, 49 \rangle$ is the same.

We examine the performance of the IC-DSA architecture compared with the OSA architecture. In the latter case, we assume PU traffic is relayed by PU nodes only, and network coding is used to encode packets whenever possible. Each PU or SU node pair uses two bidirectional routing paths, and has demand 1 (uniform demand).

Figure 3.7a plots the PU and SU throughputs in the OSA , and the IC-DSA architectures, and the win-win zone in the IC-DSA architecture. The PU and SU throughputs in the OSA architecture are obtained as follows. First we optimize the PU throughput η using the LP model in Fig. 3.2 (with λ as a free variable). Denote the obtained optimal value as η^*_{OSA}. Then we add a temporary constraint $\eta = \eta^*_{\text{OSA}}$ to the LP model and maximize λ to get the maximum SU throughput, denoted as λ^*_{OSA}. Note that the temporary constraint is always removed after each optimization so that the LP model remains as in Fig. 3.2 for next optimization.

The IC-DSA win-win zone indicates the simultaneously achievable PU and SU throughputs (η, λ) that are larger than $(\eta^*_{\text{OSA}}, \lambda^*_{\text{OSA}})$ in the OSA architecture. It is obtained as follows. First we add a temporary constraint $\lambda = \lambda^*_{\text{OSA}}$ (the maximum SU throughput of OSA) and then optimize η to get the PU throughput in the IC-DSA architecture, denoted as $\eta^*_{\text{IC-DSA}}$. Note that although we use the same LP model for both the OSA and the IC-DSA architectures, the data fed to the LP model, specifically the routing paths, coding modes, and conflict graph cliques, are different (α_k and β_k are the same for both). Next we evenly pick 10 values $t_1, \ldots, t_{10}$ from the range $[\eta^*_{\text{OSA}}, \eta^*_{\text{IC-DSA}}]$, with $t_1 = \eta^*_{\text{OSA}}$ and $t_{10} = \eta^*_{\text{IC-DSA}}$, and run 10 optimizations. For the ith optimization ($1 \leq i \leq 10$), we add a temporary constraint $\eta = t_i$ and maximize λ. The obtained optimal value of λ is denoted as λ_i^*. The line segment represented by points $\{(t_1, \lambda_1^*), \ldots, (t_{10}, \lambda_{10}^*)\}$, and the lines $x = \eta_{OSA}^*$, $y = \lambda_{OSA}^*$, together form the boundary of the win-win zone. Note that point $(\eta_{OSA}^*, \lambda_{OSA}^*)$ is the smallest point in the win-win zone. A point (η, λ) in the win-win zone indicates the PU and SU throughputs that are simultaneously achievable with IC-DSA, i.e., it is a win-win operation state for the network. The area above the top boundary indicates PU and SU throughputs not simultaneously achievable. Certainly, if either PU or SU throughput is refrained, i.e., making it smaller than η_{OSA}^* or λ_{OSA}^*, then the other can be further increased (to a certain point). Nevertheless, this is not a win-win situation and therefore, the network operation should avoid it. Note that for PU, in addition to increased throughput, the packet delay is also reduced due to shorter routing paths.

The win-win zone is important for the operation of a DSA network. Specifically, the network operator can calculate the win-win zone in the IC-DSA architecture and then increases the PU and SU throughputs from $(\eta_{OSA}^*, \lambda_{OSA}^*)$ to (η, λ), to offer better service (throughput) to the PU and the SU, as long as (η, λ) is in the win-win zone. The operator can also use weights for PU and SU traffic, denoted as w_p and w_s, to maximize revenue or other utility via optimizing $w_p \eta + w_s \lambda$, subject to (η, λ) belonging to the win-win zone, to ensure that PU and SU throughputs are not degraded compared with the one in the OSA architecture.

The win-win zone in Fig. 3.7a is large, which means that with the IC-DSA architecture, both PU and SU throughputs can increase significantly. In particular, SU throughput can increase from 0 to as high as 3.9, because the PU traffic has been rerouted to the paths illustrated in Fig. 3.6, and even though the new PU paths still cross the SU path $\langle 29, 49 \rangle$, the network coding between PU and SU traffic gives coding and transmission opportunity to SU traffic. Furthermore, the PU throughput also increases dramatically with IC-DSA, from 5 to as high as 8.33. The straight top boundary of Fig. 3.7a indicates that PU and SU throughputs form an approximately linear relationship. This is likely because the neighborhoods of nodes 38 and 40 are bottlenecks, and thus if there is more PU traffic traversing node 38 (or 40), the SU traffic transiting node 38 (or 40) on path $\langle 29, 49 \rangle$ would decrease proportionally. More specifically, at the maximum throughput, the PU traffic load α and the SU traffic load β transiting node 38 have the relationship $\alpha + \beta +$background link traffic=capacity at node 38. Therefore, the PU and SU throughputs have an approximately linear relationship. Nevertheless, this relationship is also affected by other nodes on the path, and is not obvious in some experiments with random demands (to be discussed later).

Figure 3.7b shows the win-win zone averaged on 30 experiments with random demands. The win-win zone is larger than the one in Fig. 3.7a. The SU throughput can go from 0 to as high as 6.47, and PU throughput can go from 8.26 to as high as 14.9. The win-win zone in Fig. 3.7b has a more smooth shape than the one in Fig. 3.7a, which has a sharp turn at $\eta = 6.33$. This is because the win-win zones obtained by some experiments have a smooth shape. Furthermore, the averaging operation also contributes to a smooth shape.

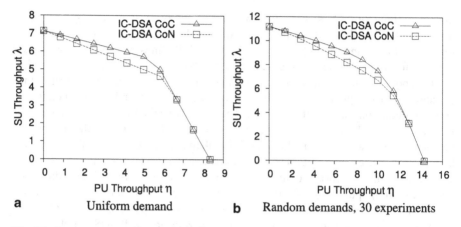

Fig. 3.8 Compare win-win zones of CoC and CoN

Next we evaluate the performance of the IC-DSA architecture using the CoC approach for network coding instead of CoN. Figure 3.8 compares the IC-DSA win-win zones obtained using CoC and CoN, respectively. The win-win zones of CoN are those in Fig. 3.7. The win-win zones of CoC are drawn using circled lines as the top and bottom boundaries. They are larger and include the entire win-win zones of CoN, i.e., CoC obtains higher throughputs. In the following, we analyze the benefit of CoC over CoN. We take the packet transmission at node 38 as an example. In Fig. 3.6, each of the two paths crossing node 38 is a scenario of *information exchange* studied in [8]. As such, no matter it is CoC or CoN, every intermediate node on each path, except node 38, will send and receive packets as coded to get higher throughput, e.g., node 31 will use coding mode [(24→31→38,'c'), (38→31→24,'c')]. Hence node 38 will receive coded packets from all neighbors, denoted as $\gamma_1 \oplus \gamma_2$ from node 31, $\gamma_3 \oplus \gamma_4$ from node 37, $\gamma_5 \oplus \gamma_6$ from node 39, and γ_7 from node 45 (which sends native packet only, since it is the end node). Suppose packets $\gamma_1, \gamma_3, \gamma_5$ go inbound to node 38 and $\gamma_2, \gamma_4, \gamma_6$ go outbound against node 38. With CoC, node 38 will directly encode all four received packets as $(\gamma_1 \oplus \gamma_2) \oplus (\gamma_3 \oplus \gamma_4) \oplus (\gamma_5 \oplus \gamma_6) \oplus \gamma_7$ to all four neighbors. With CoN, node 38 will first decode the received packets to get the intended native packets, $\gamma_1, \gamma_3, \gamma_5, \gamma_7$, and then encode them into two coded packets[4], $\gamma_1 \oplus \gamma_7$ to nodes 31 and 45, and $\gamma_3 \oplus \gamma_5$ to nodes 37 and 39. We can see that with CoC, we need 5 packet transmissions to exchange 4 packets $\gamma_1, \gamma_3, \gamma_5, \gamma_7$ across node 38, while with CoN, we need 6 transmissions. Therefore, the CoC throughput can be $\frac{6}{5} = 1.2$ times of the CoN throughput. If the traffic load on one path is more than on the other path, then the extra packets on the first path need to be transmitted separately, i.e., exchanging 2 packets using 3 transmissions, and

---

[4] Note that with CoN, node 38 will not encode the four native packets as $\gamma_1 \oplus \gamma_7 \oplus \gamma_3 \oplus \gamma_5$ to all neighbors, as this would require nodes 31, 37, 39 send $\gamma_1, \gamma_3, \gamma_5$ as native, which would then make nodes 31, 37, 39 send $\gamma_2, \gamma_4, \gamma_6$ as native too, resulting in 3 additional packet transmissions.

the benefit of CoC will decrease, since less traffic can utilize the coding mode of exchanging 4 packets using 5 transmissions. If the traffic load is quite imbalanced, the throughput of CoC will decrease to the one of CoN. This is verified in Fig. 3.8, where CoC has more benefit in the left part of the win-win zone, while in the right part, the benefit diminishes since the traffic is more and more imbalanced.

References

1. Shi Y, Hou T (2008) A distributed optimization algorithm for multi-hop cognitive radio networks. Proceedings of IEEE Infocom, pp 1292–1300
2. Jain K, Padhye J, Padmanabhan VN, Qiu L (2003) Impact of interference on multi-hop wireless network performance. In Proc. ACM MobiCom, pp 66–80
3. Sengupta S, Rayanchu S, Banerjee S (2007) An analysis of wireless network coding for unicast sessions: the case for coding-aware routing. Proceedings of IEEE Infocom, pp 1028–1036
4. Gupta P, Kumar PR (2000) The capacity of wireless networks. IEEE Trans Inf Theory 46(2):388–404
5. Katti S, Katabi D, Hu W, Rahul H, Medard M (2005) The importance of being opportunistic: practical network coding for wireless environments. Proceedings of 43rd Allerton Annual Conference on Communication, Control, and Computing
6. Katti S, Rahul H, Hu W, Katabi D, Médard M, Crowcroft J (2006) XORs in the air: practical wireless network coding. Proceedings of the ACM SIGCOMM, pp 243–254
7. Dong Q, Wu J, Hu W, Crowcroft J (2007) Practical network coding in wireless networks. In Proc. ACM MobiCom, pp 306–309
8. Wu Y, Chou P, Kung S-Y (2005) Information exchange in wireless networks with network coding and physical-layer broadcast. Proceedings of 39th Annual Conference on Information Sciences and Systems (CISS)

Chapter 4
Dynamic Spectrum Co-Access

4.1 Introduction

In the OSA architecture, whenever the PU traffic re-appears on a band, SUs must vacate from the band immediately and the on-going SU communication is disrupted. The requirement that SUs cannot access spectrum simultaneously with PUs results in significant overhead on spectrum sensing and spectrum handoff, which in turn results in poor performance for cognitive radio networks.

In this chapter, we present a dynamic spectrum sharing architecture, called *dynamic spectrum co-access* (DSCA), to resolve the disruption to secondary communications by the reappearance of PU signals. The DSCA architecture allows SUs to simultaneously access licensed spectrum with PUs, while the PU performance is not affected. Hence, unlike the OSA architecture, the SUs and PUs in the DSCA architecture coexist amicably. Generally, PUs are not willing to share spectrum with SUs. The novelty of DSCA is that the SU communication in a licensed band can provide incentives to the PU. Hence, PUs become happy to allow SUs to simultaneously access spectrum. The DSCA architecture is similar to the IC-DSA architecture introduced in the last chapter in that both enable SUs to simultaneously access spectrum with PUs through providing incentives to PUs. The difference is that in the DSCA architecture, SU transmissions are transparent to PUs. In particular, PUs do not need to decode the packets from SUs. In summary, the DSCA has the following merits.

- The data rate of PUs is ensured not to decrease, and can even increase, when SUs simultaneously access spectrum with PUs.
- PU transmitters and receivers operate without needing any knowledge of SU transmitters or receivers, and SU spectrum access. In other words, the SU spectrum access is transparent to PUs.
- Thanks to simultaneous spectrum access, the disruption to SU communications by the PU signal reappearance is eliminated.

In the DSCA architecture, a channel precoding technique, *dirty paper coding* (DPC) [1–3], is employed to achieve coexistence between PUs and SUs. Specifically, DSCA exploits DPC to enable PUs and SUs to simultaneously access spectrum, while

© The Author(s) 2015
C. Xin, M. Song, *Spectrum Sharing for Wireless Communications*,
SpringerBriefs in Electrical and Computer Engineering, DOI 10.1007/978-3-319-13803-9_4

Fig. 4.1 Coexisting on a link
with a normalized (1, a, b, 1)
channel. The legend on a link
indicates the path loss. SU is
assumed to have the PU
packet a priori

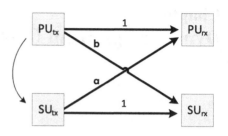

the mutual interference is eliminated, or controlled to be below the level without simultaneous spectrum access.

4.2   DSCA with One PU Node Pair and One SU Node Pair

To illustrate the idea of how the DSCA utilizes the dirty paper coding to control interference and ensure the PU performance, this section describes the DSCA architecture with one pair of PU transmitter and receiver, and one pair of SU transmitter and receiver. The DSCA architecture for multi-hop networks will be discussed in the next section. In the DSCA architecture, when there is no PU transmission, SUs freely access spectrum. When one or more PUs are transmitting, SUs still access spectrum, but provide incentives to PUs by ensuring that the PU performance is not degraded.

DPC was introduced by Costa in [1] as a proof for maintaining SINR at the receiver given the transmitter had prior knowledge of the interference state. It was shown that DPC could achieve the largest known capacity region for cognitive radio networks in a channel model with one PU node pair and one SU node pair, as long as the SU transmitter had a priori knowledge of the PU messages [4]. Several studies have also shown that SUs can coexist with PUs without degrading the PU channel capacity [3, 5, 6]. However the success of DPC in a cognitive radio network relies on the SU transmitter having a priori knowledge of the PU transmitted packet. This is a non-trivial problem. The next section will describe how the DSCA architecture utilizes the broadcast nature of wireless communications and the overhearing during the packet forwarding in multi-hop networks to solve this problem.

Next, we describe the DSCA architecture with one pair of PU transmitter and receiver, and one pair of SU transmitter and receiver, as illustrated in Fig. 4.1. In the figure, a normalized Gaussian path loss (1, a, b, 1) channel is shown between the two transmitters (one PU and one SU transmitters) and the two receivers (one PU and one SU receivers), where a and b denote the normalized path loss from the SU transmitter to the PU receiver, and from the PU transmitter to the SU receiver, respectively. The PU transmitter sends a code word X_p to the PU receiver. In this section, we assume that the SU knows the PU packet a priori. (The next section describes how the SU transmitter obtains the PU packet a priori.) To provide incentives to the PU, i.e., ensure that the PU performance is not degraded, the SU transmitter uses a portion of its power to transmit the PU code word X_p, so that the SINR at the PU receiver is boosted. Let γ $(0 < \gamma < 1)$ denote the portion of the SU power used to transmit the PU code word X_p by the SU transmitter, and $1 - \gamma$ the portion of the SU power used to transmit its own code word, denoted as $\hat{X}_s$.

Let P_p and P_s denote the transmit power of the PU and SU transmitters, respectively. Since the PU transmitter sends the code word X_p to the PU receiver, the received power at the PU receiver from the PU transmitter is $P_p = |X_p|^2$. The SU code word for the SU transmitter is deliberately generated as $X_s = \hat{X}_s + \sqrt{\frac{\gamma P_s}{P_p}} X_p$ using DPC, where $\hat{X}_s$ is the code word to carry the SU packet, and $\sqrt{\frac{\gamma P_s}{P_p}} X_p$ is the code word to carry the PU packet. This code word is constructed using random binning to ensure that the original SU code word $\hat{X}_s$ and the PU code word X_p are statistically independent. Hence, the PU receiver receives $X_p + a(\hat{X}_s + \sqrt{\frac{\gamma P_s}{P_p}} X_p)$, where $X_p + a\sqrt{\frac{\gamma P_s}{P_p}} X_p$ represents the desired code word for transporting the PU packet, and $a\hat{X}_s$ indicates the noise incurred by the SU transmission of the SU packet that is carried by code word $\hat{X}_s$, where a is the normalized path loss as illustrated in Fig. 4.1. Thus, the PU received signal power from the code words sent by both the PU transmitter and the SU transmitter can be rewritten as

$$\left(X_p + a\sqrt{\frac{\gamma P_s}{P_p}} X_p \right)^2 = \left(\sqrt{P_p} + a\sqrt{\gamma P_s} \right)^2.$$

In addition, the received noise power from the SU transmitter is $(a\hat{X}_s)^2$. Since code words $\hat{X}_s$ and X_p are statistically independent, we have $\hat{X}_s X_p = 0$. Hence, we have

$$(a\hat{X}_s)^2 = a^2 \left(X_s - \sqrt{\frac{\gamma P_s}{P_p}} X_p \right)^2 = a^2(1 - \gamma)P_s.$$

Including the normalized channel noise 1, the resulting maximum achievable rate for the PU channel, denoted as R_p, is as follows.

$$R_p = \frac{1}{2} \log \left(1 + \frac{(\sqrt{P_p} + a\sqrt{\gamma P_s})^2}{1 + a^2(1 - \gamma)P_s} \right) \tag{4.1}$$

The SU receiver receives $\hat{X}_s + \left(\sqrt{\frac{\gamma P_s}{P_p}} + b \right) X_p$, where $\hat{X}_s$ is the desired code word, and $(b + \sqrt{\frac{\gamma P_s}{P_p}})X_p$ is the interference from the PU code word X_p (transmitted by both the PU transmitter and the SU transmitter). The interference $(b + \sqrt{\frac{\gamma P_s}{P_p}})X_p$ is known by the SU transmitter in advance. Hence it can be precoded with DPC to result in an effect that this interference is eliminated at the SU receiver. Thus, the total noise at the SU receiver is the normalized channel noise 1. Thus, the SU achievable rate, denoted as R_s, is given as follows, since the received signal power $\left|\hat{X}_s\right|^2 = (1-\gamma)P_s$. (Note that the SU transmitter uses $1-\gamma$ portion of its power to transmit the SU code word.)

$$R_s = \frac{1}{2}\log\left(1 + (1-\gamma)P_s\right) \tag{4.2}$$

4.2.1 Coexistence Constraints

The coexistence constraints ensure that both PUs and SUs benefit from the coexistence. In the DSCA architecture, the SU transmitter uses a γ portion of its power to transmit the PU packet. It is important for the SU transmitter to find a proper value for γ. To offer sufficient incentives to the PU, there is a minimum value for γ. On the other hand, for the SU to obtain a certain level of performance, there is a maximum value for γ. Next we derive the constraint on the value of γ such that a win-win situation is created for both the PU and the SU, i.e., the PU obtains sufficient incentives and the SU maximizes its performance. As discussed earlier, the incentive provided to the PU is the boosted SINR at the PU receiver so that the PU performance is not degraded, but improved. Let K ($K \geq 0$) denote the required increase of the SINR for the PU in the presence of an active SU transmission. This is a measurable incentive for the PU to allow the SU to simultaneously access spectrum. Note that without the SU spectrum access, the PU achievable rate is $\frac{1}{2}\log(1 + P_p)$, where P_p indicates the SINR under the normalized channel noise 1. Thus, given the SINR incentive K to the PU, the achievable rate increases to $\frac{1}{2}\log(1 + P_p + K)$. Hence, the DSCA architecture needs to guarantee the PU achievable rate to satisfy the following relationship.

$$\frac{1}{2}\log\left(1 + \frac{(\sqrt{P_p} + a\sqrt{\gamma P_s})^2}{1 + a^2(1-\gamma)P_s}\right) \geq \frac{1}{2}\log(1 + P_p + K) \tag{4.3}$$

With some algebra manipulation, this inequality can be solved to obtain

$$\gamma \geq \left(\frac{\sqrt{(P_p + K)(1 - P_p + a^2 P_s(P_p + K + 1))} - \sqrt{P_p}}{a\sqrt{P_s}(P_p + K + 1)}\right)^2. \tag{4.4}$$

Thus, we have obtained the minimum value for γ in order to provide the SINR incentive K to the PU. Next, we find the maximum value for γ to achieve a certain level of achievable rate for the SU. Let λ denote the minimum received SINR at the SU receiver that is desired by the SU for helping the PU using a γ portion of its power. As discussed in the derivation of (4.2), in the DSCA architecture, the SINR at the SU receiver is $(1-\gamma)P_s$. Hence, to satisfy the minimum SINR λ for the SU, we must have $\lambda \leq (1-\gamma)P_s$. This is transformed into

$$\gamma \leq 1 - \frac{\lambda}{P_s}. \tag{4.5}$$

Given the constraints for the required minimum SINR for the SU, the SINR incentive for the PU, and the channel gain of the links, it is possible to determine eligible SU transmitter and receiver pairs that can coexist with a given PU transmitter and receiver pair, such that the coexistence constraints for both the PU and the SU, K and λ, are satisfied. We will introduce a concept, region of coexistence, which is a region where SUs can coexist with PUs, to generalize this in the next section.

4.3 DSCA with a Multi-Hop PU Network

We now discuss the DSCA architecture with multi-hop PU networks. There are abundant real world multihop PU networks that the DSCA architecture can be applied to. For example, terrestrial digital TV broadcasts in the United States are routinely retransmitted by both high power and low power TV translators, to help provide services to low signal areas. There is also an increasing number of wireless mesh network deployments. Other examples include emergency service networks and military networks. For all such multi-hop PU networks, the DSCA architecture can help SUs to simultaneously access spectrum with PUs, while the PUs are incentivized to allow the spectrum co-access.

In the preceding section, we have assumed that the SU transmitter knows the packet to be transmitted by the PU transmitter in advance, before both the PU and the SU transmission. The DSCA architecture utilizes the broadcast nature of wireless communications, and the overhearing in the multi-hop packet forwarding in the PU network to let SU nodes obtain the PU packets in advance. Let us consider a scenario where the PU nodes use a standard TDMA access scheme with round-robin channel access. The forwarding of a PU packet along a multi-hop path would allow the SU nodes around the path to overhear the packet. For instance, in Fig. 4.2, suppose a PU packet needs to be sent from PU_1 to PU_3 on the multi-hop path $\{PU_1, PU_2, PU_3\}$. When PU_1 transmits the packet to PU_2, SU_1 can overhear the PU packet. Afterwards, when PU_2 forwards this PU packet to PU_3, SU_1 already has the knowledge of the PU packet. Hence the SU link (SU_1, SU_2) can simultaneously access spectrum with the PU link (PU_2, PU_3) as in the preceding section, where one pair of PU transmitter and receiver co-access spectrum with one pair of SU transmitter and receiver provided that the SU transmitter knows the PU packet in advance.

For the ease of description, we use the sample topology in Fig. 4.2 to illustrate the DSCA architecture with multi-hop PU networks. In the figure, the symbol on each link denotes the path loss. The objective is to find the parameter γ (portion of power to transmit the PU packet) that each SU transmitter should use to maximize the PU achievable rate while providing sufficient incentive to the PU, given the coexistence constraints for the PU and the SU, i.e., the SINR increase K for the PU and the minimum SINR λ for the SU. The PU network is assumed to have some mechanisms to avoid mutual interference among PUs. In Fig. 4.2, PU_4 either acts like a repeater for PU_1 to forward the PU packet simultaneously as PU_2, based on the TDMA and the round-robin channel access assumption, or transmits an unrelated PU packet

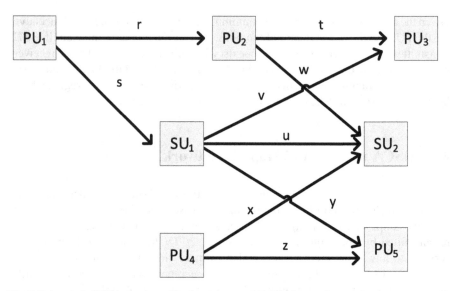

Fig. 4.2 A sample DSCA topology. The legend on each link indicates the path loss

that causes interference to PU_2. For each case we find the parameter γ for the SU transmitter SU_1 and the achievable rate on each link.

Case 1: PU_4 Transmits the Same Code Word X_P as PU_2

Given the path loss for each link shown in Fig. 4.2, and that PU_2 and PU_4 simultaneously transmit the same code word X_P, the achievable rate of link (PU_2, PU_3) is

$$R_{(PU_2,PU_3)} = \frac{1}{2} \log_2 \left(1 + \frac{(t\sqrt{P_s} + v\sqrt{\gamma P_s})^2}{N + v^2(1 - \gamma)P_s} \right),$$

where N denotes the Gaussian noise. If we consider the PU coexistence constraint K, the minimum required rate of link (PU_2, PU_3) is

$$R^*_{(PU_2,PU_3)} = \frac{1}{2} \log_2 \left(1 + \frac{(1 + K)t^2 P_p}{N} \right).$$

We can obtain the parameter γ for SU_1 by equating these two rates and solving the quadratic. The γ is given as follows.

$$\gamma = \frac{t\sqrt{P_P} \left(\sqrt{1 + \left(1 + (1 + K)\frac{t^2 P_p}{N}\right)\left(K + (1 + K)\frac{v^2 P_s}{N}\right)} - 1 \right)^2}{v\sqrt{P_s} \left(1 + (1 + K)\frac{t^2 P_p}{N}\right)} \tag{4.6}$$

The achievable rate of link (SU$_1$, SU$_2$) considering the interference from PU$_4$ is now

$$R_{(SU_1,SU_2)} = \frac{1}{2} \log_2 \left(1 + \frac{(1-\gamma)u^2 P_s}{N + x^2 P_p} \right), \tag{4.7}$$

and the overall achievable rate of link (PU$_4$, PU$_5$) is

$$R_{(PU_4,PU_5)} = \frac{1}{2} \log_2 \left(1 + \frac{(z\sqrt{P_p} + y\sqrt{P_s})^2}{N + y^2(1-\gamma)P_s} \right).$$

Link (PU$_4$, PU$_5$) would have both the SINR boost thanks to the SU transmission of PU code word X_P, but also has the increased noise due to the SU transmission of the SU code word. Nevertheless, it is reasonable to assume that the PU network does not allow the PU$_4$ and PU$_2$ transmissions to interfere each other, it is likely that the path loss y is small. Hence the interference from the SU$_1$ transmission to link (PU$_4$, PU$_5$) is also small.

Case 2: PU$_4$ Transmits a Different Code Word X_{P2}

In this case, the broadcasts of PU$_2$ and PU$_4$ are not the same code word: PU$_4$ transmits a different code word X_{P2}. Once again, it is reasonable to assume that the PU network would not allow PU$_4$ transmission to interfere the simultaneous PU$_2$ transmission on link (PU$_2$, PU$_3$). Similarly, the PU$_4$ transmission on link (PU$_4$, PU$_5$) is also not affected by a simultaneous PU$_2$ transmission. The mutually exclusive interference PU regions implies that the pass loss y should be small. The resulting achievable rate for link (PU$_2$, PU$_3$) is not affected. However the achievable rate of link (SU$_1$, SU$_2$) becomes

$$R^*_{(SU_1,SU_2)} = \frac{1}{2} \log_2 \left(1 + \frac{(1-\gamma)u^2 P_s}{N + x^2 P_{PU_4}} \right). \tag{4.8}$$

This rate is dependent on the path loss from PU$_4$ to SU$_2$. The achievable rate of link (PU$_4$, PU$_5$) is

$$R^*_{(PU_4,PU_5)} = \frac{1}{2} \log_2 \left(1 + \frac{z^2 P_{PU_4}}{N + y^2 P_s} \right). \tag{4.9}$$

This rate is also dependent on the path loss from SU$_1$ to PU$_5$. Note that here we assume that SU$_2$ does not know X_{P2}; hence SU$_2$ is unable to use DPC with its transmission and thus SU$_2$ transmission is seen by PU$_5$ as noise. However the mutual exclusivity of the interference regions of PU$_2$ and PU$_4$, and the constrained SU region of coexistence (to be discussed) implies a small path loss y. Thus (4.9) can be rewritten to

$$R^*_{(PU_4,PU_5)} \approx \frac{1}{2} \log_2 \left(1 + \frac{z^2 P_{PU_4}}{N} \right).$$

4.4   Region of Coexistence

If the PU coexistence constraint K is not able to be met by the SU transmission,
then the PU does not allow the SU to simultaneously access the licensed spectrum.
Next, we find areas within the PU network that if an SU is located within it, the
SU transmission would be able to provide sufficient incentive to the PU to allow
simultaneous spectrum access with the SU. We call such an area as the *region of
coexistence*. It is defined by two relationships. Once again, for the ease of description,
let us consider the sample topology in Fig. 4.2 and the region of coexistence for PU_2.
First, SU_1 must be able to receive the PU_1 broadcast at least as well as PU_2. This
implies the constraint

$$|r|^2 \geq |s|^2. \tag{4.10}$$

The achievable rate of link (PU_2, PU_3) is dependent only on path losses t and v.
We assume that the PU network is static (or has low mobility) and the channel is slow-
fading. Then the value of t can be considered constant during a packet transmission.
Therefore we can find a value of v that still guarantees the coexistence constraint K.
This can be obtained using (4.6) and yields

$$v = \frac{1 - \sqrt{1 - K\left(\frac{t^2 P_p}{N}(K+1)(\frac{1}{\gamma} - 1) - 1\right)}}{\frac{t\sqrt{P_p P_s}}{N\sqrt{\gamma}}(K+1)(1-\gamma) - \frac{\sqrt{\gamma P_s}}{t\sqrt{P_p}}}. \tag{4.11}$$

The constraints for r and v as indicated in (4.10) and (4.11) can be used to determine a region of coexistence for PU_2, within which the SU transmitter can safely
coexist with PU_2, i.e., simultaneously access spectrum. The values of path losses are
equivalent to two circles around PU_1 and PU_3 with radius r and v, respectively. The
overlapping region between these two circles indicates the region of coexistence for
PU_2, as illustrated in Fig. 4.3.

In Fig. 4.3, the location for the SU transmitter that can simultaneously access
spectrum with PU_2 is shown. The graph represents the potential SU achievable rate
given that there is at least 10 % SINR increase for link (PU_2, PU_3) from the SU
transmission, i.e., the PU coexistence constraint $K = 0.1$. The increasingly red
bands in the figure indicate that if the SU transmitter is closer to PU_2, then a higher
throughput can be achieved since the SU transmitter can use a smaller portion of
power to transmit the PU packet to satisfy the PU coexistence constraint K; thus
more power can be used for the transmission of its own packet. This is equivalent to
$\lambda \to 1$ since λ is directly related to the amount of power the SU transmitter uses for
its own transmission.

The relationship between the two coexistence constraints K and λ provides the
bounds for the region, where SU transmitters can be placed to simultaneously access
spectrum with the PU. This region of coexistence can thus be obtained by (4.10)
and (4.11), given K and λ. Figure 4.4 illustrates the different regions of coexistence

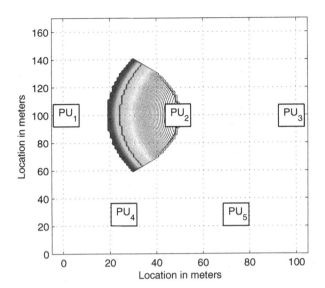

Fig. 4.3 Region of coexistence of PU$_2$

given different values of K and λ. Without a surprise, the figure indicates that the largest region of coexistence is the one when K is small and λ is large. Moreover, Fig. 4.4 also illustrates that the region of coexistence is primarily influenced by the PU coexistence constraint K (the required PU SINR increase).

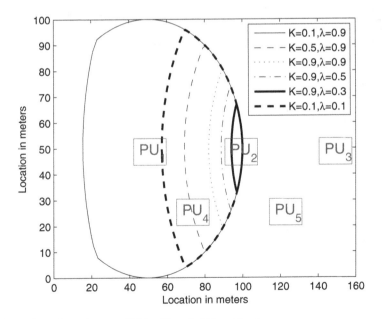

Fig. 4.4 Regions of coexistence of PU$_2$ with varied K and λ values

4.5 Coexistence Links Selection

Given the regions of coexistence of each PU node, we present an algorithm, called
Coexistent Primary and Secondary Links Selection (CoSS), to find the most beneficial
coexisting pairs of PU/SU links, to maximize the total network capacity. The PU
and the SU networks are represented as two weighted directed graphs $\mathcal{G}_1$ and $\mathcal{G}_2$,
respectively. The PU network $\mathcal{G}_1 = (\mathcal{N}, \mathcal{L})$ consists of PU node set $\mathcal{N}$ and link set
$\mathcal{L}$. The SU network $\mathcal{G}_2 = (\mathcal{M}, \mathcal{J})$ consists of SU node set $\mathcal{M}$ and link set $\mathcal{J}$.

The CoSS algorithm seeks to find the best SU link to pair with each PU link,
while satisfying the PU coexistence constraint and maximizing the SU achievable
rate. It consists of two parts. First, it selects eligible candidate SU links that satisfy
the PU coexistence constraint for each PU link. Second, it finds the SU link with the
highest achievable rate amongst the eligible candidate SU links for a PU link. The
eligible SU links are selected by applying constraint (4.10) onto all SU links that
satisfy this constraint for a given PU link l. These eligible SU links are placed in sets
C_l for potential coexistence with PU link l. Then in Part 2 the γ value for each SU
link in set C_l is obtained based on (4.4). Since this value is directly related to the SU
maximum achievable rate based on (4.7) and (4.8), the candidate SU links can be
sorted by their achievable rate. At last, the SU link with the highest achievable rate is
chosen to coexist with the given PU link. After the CoSS algorithm terminates, the
best SU coexistence link for each PU link is found, and the output is a set of PU/SU
links pairs that are the best coexistence pairs to maximize network performance.

Algorithm 1: Coexistent Primary and Secondary Links Selection (CoSS)

    Input: INPUT: Graphs $\mathcal{G}_1 = (\mathcal{N}, \mathcal{L})$ and $\mathcal{G}_2 = (\mathcal{M}, \mathcal{J})$
    Output: OUTPUT: S is a set containing PU/SU pairs of coexistent links
1 $C_l = \emptyset, S = \emptyset$
2 //Part 1
3 **for** $l \in \mathcal{L}$ **do**
4   **for** $j \in \mathcal{J}$ **do**
5        $z_{l,j}$ = path loss from link $l.head$ to link $k.head$
6   **if** $weight(l) \leq z_{l,j}$ **then**
7 | $C_l = C_l \cup \{j\}$
8   **end**
9   **end**
10 **end**
11 // Part 2
12 **for** $l \in \mathcal{L}$ **do**
13   **for** $y \in C_l$ **do**
14 | Calculate $\gamma_{l,y}$ by (4.6)
15   **end**
16    $s = \text{argmin}_{y \in C_l} \gamma_{l,y}$
17   **if** $0 \leq s \leq 1 - \frac{\lambda}{P_c}$ **then**
18 | $S = S \cup (l, s)$
19   **end**
20 **end**

Fig. 4.5 Performance of coexistent link (SU$_1$, SU$_2$)

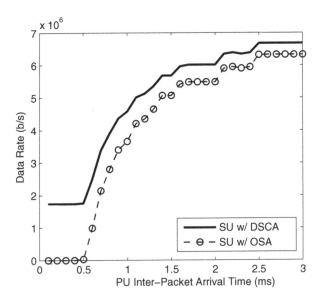

4.6 Performance Evaluation

We evaluate the performance of the DSCA architecture, and compare it with the OSA architecture under variable traffic conditions. In the OSA architecture, SUs attempt to access the gaps in the time domain between PU transmissions. We assume that the SU can accurately detect the end of a PU transmission, and then access the unused band until the PU returns. We use two sample topologies, a small topology illustrated in Fig. 4.2 and a large topology with a PU cellular network.

For the small topology in Fig. 4.2, we assume one PU multicast flow from PU$_1$ to PU$_3$ and PU$_5$, with PU$_2$ and PU$_4$ as the relay nodes. The packet arrival is assumed to follow the Poisson process, with the mean inter-packet arrival time denoted as ρ. The packet size is assumed 500 bytes. To focus on evaluation of the coexistence between the PU and the SU, we assume that SU$_1$ always has traffic to SU$_2$, which eliminates the impact of the SU traffic load on the performance. The resulted SU throughput with such backlogged traffic assumption is called *saturation throughput*, which is approximately the system capacity [7]. The distance of all PU and SU links is assumed 50 m. We assume a 20 MHz channel with the transmit power of 500 mW and a path loss of d^{-3}, where d is the distance between two nodes. The PU ad SU coexistence constraints are $K = 0.5$ and $\lambda = 0.9$, respectively. In the OSA architecture, we assume that at least 1 DIFS of 50 ns is required for the SU to detect an idle channel, while 10 DIFS = 0.5 ms is required for the SU to detect the PU activity on the channel.

The achievable rates of links (PU$_2$, PU$_3$) and (SU$_1$, SU$_2$) are shown in Figs. 4.5 and 4.6, respectively. In Fig. 4.5, the advantage of the DSCA architecture is clear when the PU network is saturated with traffic, i.e., $\rho \to 0$. Since the PU network

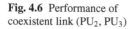

Fig. 4.6 Performance of coexistent link (PU_2, PU_3)

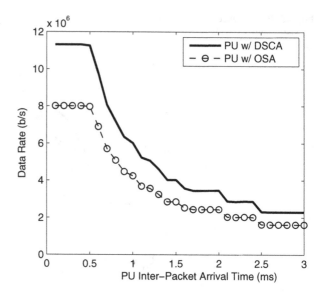

is using every available network channel access, the SUs are unable to gain channel access in the OSA architecture. However, the DSCA architecture is able to exploit simultaneous transmissions between the SU and the PU. As the packet inter-arrival time increases, the performance gap of the two architectures becomes smaller, since the simultaneous transmission becomes less needed.

When $\rho < 0.4$, the SUs under the DSCA architecture achieve a baseline performance of approximately 1.8×10^6 b/s. This indicates that the SUs with the DSCA architecture are able to communicate at some minimum rate regardless of PU activity, i.e., SUs can simultaneously access spectrum with PUs to obtain some minimum rate.

The impact of the SU transmission on the PU transmissions can be seen in Fig. 4.6. It is clear that the DSCA architecture provides a higher PU rate for all levels of PU traffic. In summary, the DSCA architecture is able to provide the desired SINR increase for the PU as well as significantly improve the SU performance.

Next we use a topology with 37 PU nodes deployed in a hexagonal cellular grid with 100 m on each edge, in an area of 600×600 m. The PU node at the center of the area is called the *gateway*. In the network, the distance from the gateway to any other node in the network is at most four hops. The gateway periodically broadcasts packets to all the nodes in the network as follows. A packet is first transmitted by the gateway, and received by the PU nodes with one-hop distance to the gateway. Then the one-hop PU nodes from the gateway forward the packet to the nodes with two-hop distance to the gateway. Similarly, the two-hop nodes from the gateway forward the packet to the nodes with three-hop distance to the gateway. Finally, the three-hop nodes from the gateway forward the packet to the nodes with four-hop distance to the gateway. As a result, the entire network has received the packet. The PU nodes is equipped with omnidirectional antennae with 0.5 W power limitation. The SU

Fig. 4.7 Percentage of links involved in coexistence in the large topology

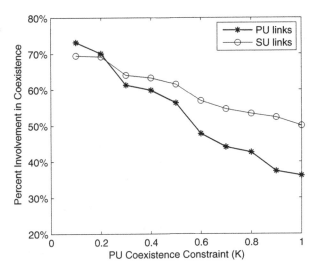

Fig. 4.8 PU network performance increase in the large topology

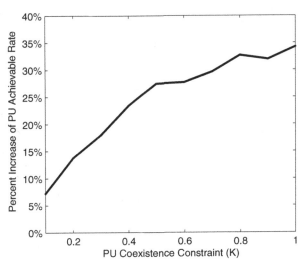

network has 30 nodes randomly deployed within the PU network area. Similar to the preceding subsection, we also assume each SU node always has outgoing traffic. The transmission radius of each SU node is assumed 100 m. The SU nodes are also equipped with omnidirectional antennae with 0.5 W power limitation.

The results for the large topology are shown in Figs. 4.7 and 4.8. The PU coexistence constraint K is set from 0.1 to 1, i.e., 10 to 100 % SINR increase for the PU. In Fig. 4.7, when the PU coexistence constraint $K = 0.1$, over 70 % of PU links coexist with SU links. This number is significant as the gateway and the one-hop nodes from the gateway are not eligible for coexistence with SU links. Thus on average, 25 of the eligible 30 PU nodes benefit from coexistence with SU nodes. Although

Fig. 4.9 Percent of PU links
involved in coexistence,
$\gamma = 0.5$

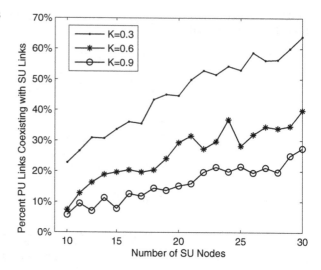

Fig. 4.10 Percent of PU links
involved in coexistence,
$K = 0.1$

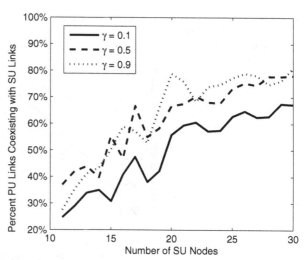

the percentage of coexisting PU links decreases as the PU coexistence constraint K
increases, there are still nearly 35 % or 11 PU nodes benefiting from coexistence
when even $K = 1$, a full doubling of the PU SINR by coexistence. The percentage
of participating SU nodes is almost 70 % when K is relatively small.

Figure 4.8 plots the overall network performance improvement with the DSCA
architecture. The PU achievable rate increases by 8 %, when the coexistence constraint $K = 0.1$. As K increases, the PU achievable rate also increases. However,
when K is large, the PU achievable rate increases slowlier, as fewer SU links qualify
for coexistence.

Next we examine the number of PU links that can benefit from SU coexistence. Figures 4.9 and 4.10 plots the percentage of PU links that simultaneously access spectrum with SU links, as a function of the number of randomly placed SU nodes. Figure 4.9 illustrates that a smaller PU coexistence constraint K results in more participation in coexistence by PUs nodes. As the number of SU nodes is close to the number of PU nodes, $K = 0.3$ results in over 63 % of PU nodes participating in coexistence. In general, a low coexistence constraint results in a significant number of PU nodes able to benefit from coexistence, which in turn increases the overall PU network throughput.

Figure 4.10 illustrates that increasing γ, the portion of SU power for transmitting PU packets, also results in higher PU participation in coexistence. However, requiring more SU power to help PU transmission does not directly translate to higher PU participation. For instance, the PU participation is similar at $\gamma = 0.5$ and $\gamma = 0.9$, which indicates diminishing returns on increasing γ to improve PU participation.

From Figs. 4.9 and 4.10, one conclusion that can be drawn is that PUs have significant control over how many of their nodes benefit from coexistence by adjusting the coexistence constraint K.

References

1. Costa M (1983) Writing on dirty paper (corresp.). IEEE Trans Inf Theory 29(3):439–441
2. Goldsmith A, Jafar SA, Maric I, Srinivasa S (2009) Breaking spectrum gridlock with cognitive radios: an information theoretic perspective. Proc IEEE 97(5):894–914
3. Jovicic A, Viswanath P (2009) Cognitive radio: an information-theoretic perspective. IEEE Trans Inf Theory 55(9):3945–3958
4. Devroye N, Mitran P, Tarokh V (2006) Achievable rates in cognitive radio channels. IEEE Trans Inf Theory 52(5):1813–1827
5. Guan X, Cai Y, Sheng Y, Yang W (2012) Exploiting primary retransmission to improve secondary throughput by cognitive relaying with best-relay selection. IET Commun 6(12):1769–1780
6. Naganawa J, Kobayashi K, Katayama M, Yamazato T (2010) Performance evaluation of the cognitive piggyback overlay systems with dirty paper coding. In: Communications and Information Technologies (ISCIT), 2010 International Symposium on. IEEE, pp 974–979
7. Bianchi G (2000) Performance analysis of the IEEE 802.11 distributed coordination function. IEEE J Sel Areas Commun 18(3):535–547

Chapter 5
On-Demand Spectrum Access

5.1 Introduction

In this chapter, we present an application-oriented spectrum sharing architecture, called *on-demand spectrum access* (ODSA), to eliminate the technical barriers bothering the OSA architecture, so that users can dynamically and efficiently share spectrum in a well-managed and streamlined architecture. In the ODSA architecture, a spectrum service provider offers on-demand spectrum services to users, such that users can dynamically set up *application-oriented virtual topologies* to carry out specific applications. For instance, a virtual topology can be set up among a set of nodes to transport a large data flow or a video conference. Enabling users to set up application-oriented virtual topologies is a unique feature of the ODSA architecture, as this directly targets the performance experienced by end users. As an additional merit, the ODSA architecture provides guaranteed quality of service (QoS) for users through dedicated spectrum allocation for virtual topologies, while the spectrum is still efficiently shared through dynamic spectrum services.

The ODSA architecture is different from the OSA architecture on both technical complexity and easiness for management. The time scale for the latter is typically at the order of packet transmission time, while the time scale for the ODSA architecture is at the order of application session duration, which is typically from minutes to days. Moreover, there is no distinction between PUs and SUs and hence spectrum sensing is not needed in the ODSA architecture. As a result, the technical complexity of the ODSA architecture is significantly reduced and the management is significantly streamlined, compared with the OSA architecture.

In the literature, a related DSA approach called *dynamic spectrum allocation* or *coordinated dynamic spectrum access* has been studied [1, 2]. These studies considered how to allocate spectrum for base stations in a cellular network, to maximize the number of base stations to get their requested spectrum, subject to interference constraint. In [1], this problem was formulated as an *integer programming* problem. In [2], a heuristic algorithm based on Tabu search was developed. Unlike these studies, the ODSA architecture considers multi-hop spectrum allocation to satisfy an end-to-end spectrum demand in a mesh network. In the ODSA architecture, the entity for spectrum allocation is not a single node (a base station), but a point-to-point

© The Author(s) 2015
C. Xin, M. Song, *Spectrum Sharing for Wireless Communications,*
SpringerBriefs in Electrical and Computer Engineering, DOI 10.1007/978-3-319-13803-9_5

Fig. 5.1 Spectrum service provider infrastructure mesh network, and the virtual topologies established by two spectrum demands: (1) demand (A, B) to set up a P2P path between nodes A and B; and (2) demand (C, D, E) to set up a P2MP path among nodes C, D, and E

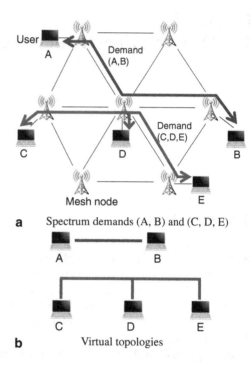

a   Spectrum demands (A, B) and (C, D, E)

b   Virtual topologies

or point-to-multipoint path, to carry out an application such as a video conference. Moreover, the constraints for spectrum allocation in ODSA can be quite different from the ones in [1, 2]. For the latter, two adjacent base stations cannot be allocated with the same band, because of the interference constraint. The constraints for spectrum allocation in ODSA are application-oriented. For example, certain spectrum services need the same band on all nodes along the path, such as network coding, to exploit the broadcast nature of wireless communications.

5.2   Spectrum Service

In the ODSA architecture, the spectrum is managed by a spectrum service provider. For example, FCC and/or NTIA may designate an agent to manage the 500 MHz spectrum of the National Broadband Plan. This agent is seen as a spectrum service provider in the ODSA architecture. The spectrum service provider has a mesh infrastructure network, as illustrated in Fig. 5.1a. All mesh nodes are equipped with multiple cognitive radios or one cognitive radio with multiple transceivers, such as the WNaN cognitive radio [3]. The users are wirelessly connected to the mesh nodes, and send spectrum demands to the spectrum service provider for requesting services. For the ease of description, in this chapter, a *node* refers to the mesh node in the mesh network of the spectrum service provider, and a *user* refers to a user terminal connected to a mesh node of the spectrum service provider infrastructure network. Based on the service type requested by a user demand, the spectrum service provider sets up a corresponding virtual topology among the users specified in the demand, to carry out an application, e.g., deliver a large data file or a video conference. A virtual topology is set up by allocating the required number of spectrum band(s) requested by the user on each mesh node along a *point-to-point* (P2P) or *point-to-multipoint* (P2MP) path connecting these users, and switching the cognitive radios (or transceivers) of these mesh nodes to the allocated band(s). Note that multiple virtual topologies can be supported simultaneously since each cognitive radio has multiple transceivers. Figure 5.1 illustrates the infrastructure mesh network of a spectrum service provider, and the virtual topologies set up by two spectrum demands.

The spectrum managed by the spectrum service provider is divided into multiple bands. Let M denote the total number of bands. A user demand requests one or multiple bands. Hence, in this chapter, we refer to spectrum allocation and *band allocation* interchangeably. The spectrum service can be characterized by different parameters, e.g., the type of path to support the service, bandwidth heterogeneity, interference exploitation or mitigation, and non-channelization, to accommodate diversified requirements of user applications. First, some applications may be supported by a P2P path (of mesh nodes), e.g., data transfer between two users, while others may need a P2MP path, e.g., a multicast application. Note that a P2MP path is equivalent to a multipoint-to-multipoint path with regard to band allocation; hence a multipoint-to-multipoint application can also be supported by a P2MP path. Second, we may require different demands to request the same number of bands, or allow them to request different number of bands. In the case of homogeneous band requirement, we divide the spectrum such that one band is the spectrum amount requested by each demand. Third, thanks to the broadcast nature of wireless communications, the transmissions of nearby nodes on a path result in co-channel interference. A user demand may want to mitigate co-channel interference by requesting different bands for nearby nodes on the path. For instance, if the user application is a unidirectional file transfer on a P2P path, the broadcast nature of wireless communications cannot be exploited, and hence the co-channel interference should be eliminated to improve performance. On the other hand, some applications, such as video conferencing or network coding, may exploit the broadcast nature, and hence the user demand would like all nodes on the path to be allocated with the same band(s). At last, a demand may want to request an arbitrary amount of spectrum, i.e., the spectrum is non-channelized. The non-channelized spectrum demand can be modeled by the demand with heterogeneous band requirement, as long as the spectrum is divided into bands of sufficient small width. Hence we do not need to deal with non-channelized spectrum demands separately. Based on these discussions, the spectrum services can be classified into the following eight types:

1. PP-SB-IE: point-to-point single-band interference-exploitation,
2. PP-SB-IM: point-to-point single-band interference-mitigation,
3. PP-MB-IE: point-to-point multiple-bands interference-exploitation,
4. PP-MB-IM: point-to-point multiple-bands interference-mitigation,

5. **MP-SB-IE:** point-to-multipoint single-band interference-exploitation,
6. **MP-SB-IM:** point-to-multipoint single-band interference-mitigation,
7. **MP-MB-IE:** point-to-multipoint multiple-bands interference-exploitation, and
8. **MP-MB-IM:** point-to-multipoint multiple-bands interference-mitigation.

In the ODSA architecture, a fundamental problem is band allocation for user demands, i.e., for an incoming user demand, how to allocate spectrum bands to the nodes on the P2P or P2MP path computed for the demand, to set up the application-oriented virtual topology for users. To efficiently share spectrum, user demands are allowed to dynamically arrive and depart; hence online algorithms are needed for band allocation to quickly set up the virtual topology. For an arriving demand, the spectrum service provider uses a certain routing algorithm, e.g., the shortest path algorithm, to compute a P2P or P2MP path in the spectrum service provider infrastructure network to connect the users specified by the demand. The band allocation for some spectrum services is similar. For example, services MP-SB-IE and PP-SB-IE both have the objective to find the same band among all the nodes on a P2P path (for PP-SB-IE) or a P2MP path (for MP-SB-IE). When finding a common band, both the P2P and the P2MP path can be treated as a set of nodes. Therefore, the band allocation for these two services is the same. The same is true for services MP-MB-IE and PP-MB-IE.

The band allocation can be carried out in either a *centralized* or a *distributed* mode. In the centralized mode, all user demands are sent to a central controller through a control channel. The central controller processes the user demand, conducts band allocation, and notifies the corresponding mesh nodes along the path to support the user demand. In the distributed mode, the source node to which a user is connected processes the spectrum demand from this user. It computes the path for the demand and sends a control packet along the path. Each intermediate node on the path records its free bands into the control packet, and forwards the packet to the downstream node. After the control packet arrives at the destination, it includes the free bands on all nodes along the path. The destination then conducts band allocation for the demand, and notifies all the nodes along the path to configure the cognitive radios to the allocated band(s). If it is a P2MP path, the destinations send the collected free bands information back to the source node, which finally conducts band allocation and notifies the nodes on the path.

A relevant problem to band allocation is the channel assignment (also known as link scheduling) for wireless networks. In the literature, this problem is typically reduced to graph coloring, such as strong edge coloring, or distance-2 edge coloring [4–7] . That is, two edges within distance 2 cannot use the same channel. Here two edges of distance 1 refer to adjacent edges, and two edges of distance 2 mean that there is another edge in between. The band allocation for the ODSA architecture is different from this problem on two aspects. First of all, the constraint is different. The band allocation for IE services (PP-SB-IE, PP-MB-IE, MP-SB-IE, MP-MB-IE) actually require the band for all edges to be the same. The band allocation for IM services (PP-SB-IM, PP-MB-IM, MP-SB-IM, MP-MB-IM) does require the distance-2 edges to have different band, but does not require the distance-1 edges to have different bands. Second, the demands for band allocation in the ODSA architecture dynamically/randomly arrive, one by one. The demands for channel assignment, i.e., the links information, are all known in advance. This implies that in channel assignment, all channels are available for assignment at each edge, while for band allocation in the ODSA architecture, the availability of bands dynamically change and is usually heterogeneous for different edges, which makes the allocation more complicated.

In the remainder of the chapter, we present the band allocation for two fundamental spectrum services, the PP-SB-IE and PP-SB-IM services, in the ODSA architecture. The study on band allocation for other services is still ongoing.

5.3   Band Allocation for PP-SB-IE Service

A demand for PP-SB-IE service requests a common free band on a path. We call a demand to request a spectrum band on a path between nodes s and d as a class-(s, d) demand. The band allocation for a PP-SB-IE demand is illustrated in Algorithm 2. The spectrum service provider allocates a band for an incoming class-(s, d) demand if and only if there is at least one common free band on a path between s and d. Otherwise, this demand is rejected. If there are more than one common free band on the path, then the *random band allocation* scheme is adopted, i.e., one band is randomly picked from the free bands pool.

Next, we analyze the blocking probability of the band allocation algorithm for the PP-SB-IE service. We assume fixed routing to compute the path for a demand, i.e., a demand is routed on one fixed path. As a general practice in analysis, the demands are assumed following Poisson arrival and exponential durations. Let p_{sd} denote the path between nodes s and d. One may have a concern that the nodes close to path p_{sd} may cause co-channel interference to path p_{sd}. With the random band allocation scheme, a path traversing those adjacent nodes has a low probability to be allocated with the same band as the demands on p_{sd}; hence the co-channel interference is expected to be small. Moreover, if it is necessary to completely eliminate co-channel interference from nodes close to a path, we can add these nodes to be part of the path in band allocation; the same band on those nodes will not be allocated to another demand. Table 5.1 lists the main notations to be used in this section.

First of all, we derive the probability mass function (pmf) for the number of common free bands among n nodes on a path p. Without loss of generality, we number the n nodes and refer to them as node 1, node 2, ..., and node n. Let Z_v denote the random variable indicating the number of free bands at node v. We assume that the number of free bands at different nodes are independent, which is reasonable for a mesh network. We first consider the case with only two nodes, i.e., $n = 2$. Let $\varphi(k \mid i, j)$ denote the probability that there are k common free bands given that the number of free bands at the two nodes are i and j, respectively, i.e., $Z_1 = i$ and $Z_2 = j$. Due to the random band allocation, the probability that there are k common free bands is equivalent to the probability that when randomly distributing i red balls

---

Algorithm 2: Band allocation for PP-SB-IE service

Input: Source and destination s, d of the user demand

1 Let $\mathscr{P}_{sd}$ denote the set of paths between the source node s and the destination node d that are pre-computed or computed in real time

2 **for** $p \in \mathscr{P}_{sd}$ **do**

3   Collect the set of bands $\mathscr{F}$ that are commonly free on all nodes of path p, either from a band usage database (centralized mode), or through a signaling process on the path (distributed mode)

4   **if** $\mathscr{F}$ *is not empty* **then**

5   Randomly select a band from $\mathscr{F}$ for the demand

6   Send a control packet to all nodes of path p to tune one transceiver of the cognitive radio to the selected band

7   Return **success**

8   **end**

9 **end**

10 Return **failure**

Table 5.1 Main notations

p_{sd}	Routing path between nodes s and d
λ_{sd}	Demand arrival rate between nodes s and d
Z_v	Random variable indicating the number of free bands at node v
M	Number of bands
$\pi_v(i)$	Probability that node v has i free bands
γ^p	Blocking probability of path p
$\rho_v(i)$	Arrival rate of the carried demands at node v

into M boxes, up to one ball per box, and then randomly distributing j black balls into the M boxes, also up to one ball per box, there are k boxes containing one red and one black balls. Therefore $\varphi(k \mid i, j)$ is obtained as

$$\varphi(k \mid i, j) = \begin{cases} \binom{i}{k}\binom{M-i}{j-k}/\binom{M}{j}, & \text{if } \max(i + j - M, 0) \le k \le \min(i, j) \\ 0, & \text{otherwise} \end{cases} . \qquad (5.1)$$

Next we consider the case with more than two nodes, i.e., $n > 2$. Let $\beta(k \mid i_1, \ldots, i_j)$ denote the probability that there are k common free bands among nodes 1 to j, given that $Z_v = i_v$ for $1 \le v \le j$, i.e., the number of free bands at node v is i_v. Clearly $\beta(k \mid i_1, i_2) = \varphi(k \mid i_1, i_2)$. If $n > 2$, $\beta(m \mid i_1, \ldots, i_n)$ is recursively obtained as

$$\beta(k \mid i_1, \ldots, i_n) = \sum_{m=k}^{M} \varphi(k \mid m, i_n)\beta(m \mid i_1, \ldots, i_{n-1}). \qquad (5.2)$$

Let γ^p denote the blocking probability of path p that contains nodes 1 to n. Let $\pi_v(i)$ denote that probability that there are i free bands at node v ($1 \le v \le n$). The

γ^p is given as

$$\gamma^p = \sum\nolimits_{i_1=0}^{M} \cdots \sum\nolimits_{i_n=0}^{M} \beta(0 \mid i_1, \ldots, i_n) \pi_1(i_1) \cdots \pi_n(i_n), \qquad (5.3)$$

where $\beta(0 \mid i_1, \ldots, i_n)$ indicates the conditional path blocking probability given that the number of free bands at the n nodes are i_1, ..., and i_n, respectively.

Let $\gamma_v^p(k)$ denote the blocking probability of path p given that $Z_v = k$, which can be obtained from Eq. (5.3) by restricting the value of $i_v = k$, and removing $\pi_v(i_v)$.

Now we derive the probability distribution $\pi_v(\cdot)$ for random variable Z_v, the number of free bands at node v. The demand arrival rate for path p_{sd}, denoted as $\lambda_{p_{sd}}$, is equal to λ_{sd}, the demand arrival rate between nodes s and d, since we have assumed fixed routing. Let $\mathcal{P}_v$ denote all paths that contain node v. Let $\rho_v(i)$ denote the arrival rate of the carried demands that arrive at node v and are successfully allocated with bands, given that $Z_v = i$. Then we have

$$\rho_v(i) = \sum_{p \in \mathcal{P}_v} \lambda_p \left(1 - \gamma_v^p(i)\right).$$

The band allocation at node v can be modeled as a birth-death process with $M+1$ states. State i ($0 \le i \le M$) indicates the number of free bands at node v. The birth rate at state i is $M - i$, i.e., the service rate of the $M - i$ demands carried at node v. The death rate at state i is $\rho_v(i)$ for $i > 0$, the arrival rate of the carried demands given that $Z_v = i$. The death rate at state 0 is 0. The stationary probability of the process at state i is $\pi_v(i)$, which can be obtained as follows.

$$\pi_v(i) = \pi_v(M) \prod\nolimits_{k=i+1}^{M} \rho_v(k)/(M - i)! \qquad (5.4)$$

With the constraint $\sum_{i=0}^{M} \pi_v(i) = 1$, we can solve (5.4) to obtain $\pi_v(i)$ for all $0 \le i \le M$ and $1 \le v \le n$. The equations in (5.3–5.4) form a system of non-linear equations. We can solve it to obtain γ^p for all p. Then we can compute the average blocking probability for PP-SB-IE service as

$$\bar{\gamma} = \sum_{s,d \in \mathcal{V}, s \ne d} \lambda_{sd} \gamma^{p_{sd}} / \sum_{s,d \in \mathcal{V}, s \ne d} \lambda_{sd},$$

where $\mathcal{V}$ denotes the nodes set.

5.4 Band Allocation for PP-SB-IM Service

A PP-SB-IM demand requires that the band allocation on a path is performed such that the co-channel interference between adjacent nodes is eliminated. We assume a protocol interference model similar to the one used by IEEE 802.11, i.e., if two nodes are communicating with each other, then the neighbor nodes of both the transmitter

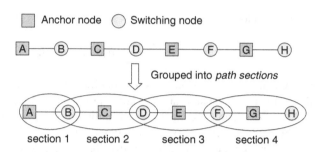

Fig. 5.2 The anchor and switching nodes on a path

and receiver are interfered and cannot send/or receive packets. This objective can be achieved by combining band allocation and transmission scheduling.

Next we present an effective and efficient band allocation algorithm for PP-SB-IM service. The nodes on a path are classified into *anchor nodes* and *switching nodes* as follows. The odd numbered nodes on the path are anchor nodes and the even numbered nodes on the path are switching nodes. When transmitting data along the path, the anchor nodes stay on their allocated bands, while the switching nodes alternatively switch between the two bands of the two adjacent anchor nodes. We call an anchor node and its immediately upstream and downstream switching nodes on the path (if any) as a *path section*, as illustrated in Fig. 5.2. A free band of a path section refers to a common free band among all nodes of this path section. We refer to a path section that has only one free band as a *single free band* (SFB) path section.

Definition 5.1 There exists a band allocation for a PP-SB-IM demand if on the corresponding path, each path section can be allocated with one free band, and the allocated bands of two adjacent path sections are different. Within a path section, the anchor node is allocated with the same band of this section, and a switching node is allocated with the two bands of the two path sections containing it.

We assume that the cognitive radios of all nodes are programmed to receive the GPS signal to be synchronized in time. The nodes are scheduled to operate in time slotted mode. The nodes on a path are scheduled to transmit as follows. An anchor node stays on its allocated band all the time. A switching node switches band every time slot as follows. At the odd numbered time slot, it switches to the band of the upstream anchor node; at the even numbered time slot, it switches to the band of the downstream anchor node. With such scheduling of node transmissions and band switching, and a band allocation by Definition 5.1, we can show that there will be no co-channel interference among the nodes along the path, under the protocol interference model.

Without loss of generality, we consider three contiguous path sections on a path, as illustrated in Fig. 5.3. Note that the following discussion is still valid if either the upstream or the downstream path sections, or both do not exist. Suppose the bands allocated to the three sections are bands a, c, and a, respectively. Note that two non contiguous path sections can be allocated with the same band, without violating Definition 5.1. Moreover, if the band for the third path section is different from the other two sections, our proof still holds. Based on the above scheduling, the nodes

Fig. 5.3 Nodes transmission
scheduling in alternating slots

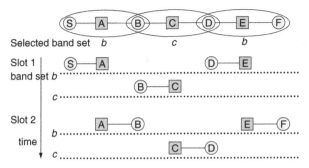

in the path sections form into communication pairs in each slot. The communication pairs and their bands in the first two slots are illustrated in Fig. 5.3. From the figure, we can see that a node does not interfere the nodes within two hop distance in the first two time slots. Furthermore, the scheduling of odd numbered slots is the same, and the scheduling of even numbered slots is also the same. In other words, the scheduling in slots 1 and 2 are repeated in slots 3 and 4, and so on and so forth. Hence, there is no co-channel interference among the nodes along the path under the protocol interference model.

In the following discussions, *we assume that every section of a path has at least one free band*. (Otherwise, an incoming PP-SB-IM demand is simply blocked.) A path section that has only one free band is called a *single free band* (SFB) path section. The path sections of a path are grouped into *path segments* with the SFB sections as boundaries. Generally, a path segment contains two adjacent SFB sections and all non-SFB sections between them. A non-SFB section has more than one free band. The first path segment starts from the first section to the first SFB section, while the last path segment is from the last SFB section to the last path section. If there is no SFB section, then the entire path is one segment. For example, suppose a path has 5 sections, $R_1, \ldots, R_5$, and R_2, R_4 are SFB sections. Then $R_1, \ldots, R_5$ are grouped into three segments: $\{R_1, R_2\}$, $\{R_2, R_3, R_4\}$, and $\{R_4, R_5\}$. The SFB path section is important for band allocation as band allocation for each path segment can be independently conducted by the following theorem.

Theorem 5.1 *If there exists a band allocation for each path segment, then there exists a band allocation for the entire path, which is the combination of the band allocations for all the path segments.*

Proof The boundary sections between path segments are SFB sections. An SFB section has only one choice for band allocation since it has only one free band. Thus the band allocation for the internal non-SFB sections inside a path segment is only affected by the boundary SFB sections, which is already fixed, and is not affected by band allocation of another path segment. Therefore, the band allocation of different path segments can be independently carried out. This proves the theorem.

With Theorem 5.1, we only need to focus on band allocation for a path segment. Next we discuss band allocation for a path segment with no SFB section, or only one SFB section (which is an end section). Assume that the path segment has n sections, denoted as $\{R_1, \ldots, R_n\}$. Let $\mathcal{B}_i$ denote the set of free bands of section R_i.

Theorem 5.2 *If a path segment contains at most one SFB section, then there exists a band allocation for the segment.*

Remark: If the path segment contains one SFB, then the SFB section must be either the beginning or the end section of the path segment.

Proof We can use Algorithm 3 to allocate a band for each path section. If the path segment contains one SFB section and it is the end section, then for the ease of description, we reverse the order of the path sections to become $\{R'_1, \ldots, R'_n\}$, where $R'_1 = R_n, \ldots, R'_n = R_1$. The algorithm sequentially scans each path section, randomly picks one band b_i from $\mathcal{B}_i$ for path section R_i, and update $\mathcal{B}_{i+1}$ by removing b_i from $\mathcal{B}_{i+1}$. The updating of $\mathcal{B}_{i+1}$ guarantees that the allocated band for a path section is different from the one allocated for the upstream path section, satisfying Definition 5.1. $\mathcal{B}_1$ contains at least one band; hence R_1 can be allocated with one band. Moreover, the updated $\mathcal{B}_{i+1}$ ($1 \leq i \leq n - 1$) is not empty since the original $\mathcal{B}_{i+1}$ contains at least two bands; hence R_{i+1} can also be allocated with a band. Therefore, there exists a band allocation for the segment.

Next we present a band allocation algorithm for a general path segment, as illustrated in Algorithm 4. The algorithm uses a forward scan from the beginning section, and a backward scan from the end section, to allocate band for each section on a segment. When allocating band for a section, if it has several choices, then it randomly picks one band. After the forward and backward scans stop, if there are still some sections not allocated with bands, then these sections must each contain at least two bands, and Algorithm 3 is used to finish band allocation for these sections. If there exists a possible band allocation, Algorithm 4 is guaranteed to find it by Theorem 5.3.

Theorem 5.3 *There is no band allocation for a general path segment $\{R_1, \ldots, R_n\}$ iff $i = n$ and $|\mathcal{B}_i| = 0$ when Algorithm 4 terminates, where n is the number of sections of the path segment.*

Algorithm 3: Band allocation for a path segment $\{R_1, \ldots, R_n\}$, with at most one SFB section

---

Input: either $\mathcal{B}_1$ or $\mathcal{B}_n$ may contain one band. All other sections contain more than one band.

1 **if** $\mathcal{B}_n = 1$ **then**
2 | Reverse the order of path sections such that R_n becomes the first section, R_{n-1} is the second section, ..., and R_1 is the nth section.
3 **end**
4 $i = 1$
5 **for** $1 \leq i \leq n$ **do**
6 | Randomly pick a band in $\mathcal{B}_i$, denoted as b_i, to section R_i
7 | **if** $i < n$ **then**
8 | | $\mathcal{B}_{i+1} = \mathcal{B}_{i+1} \setminus b_i$
9 | **end**
10 **end**
11 Return $b_1, \ldots, b_n$

Algorithm 4: Band allocation for a path segment $\{R_1, \ldots, R_n\}$

---

1 $i = 1$
2 **while** $i \leq n$ *and* $|\mathcal{B}_i| = 1$ **do**
3 | Allocate the only band in $\mathcal{B}_i$, denoted as b_i, to section R_i
4 | **if** $i < n$ **then**
5 | | $\mathcal{B}_{i+1} = \mathcal{B}_{i+1} \backslash b_i$
6 | **end**
7 | $i = i + 1$
8 **end**
9 $j = n$
10 **if** $i < n$ **then**
11 | **while** $j \geq i$ *and* $|\mathcal{B}_j| = 1$ **do**
12 | | Allocate the only band in $\mathcal{B}_j$, denoted as b_j, to section R_j
13 | | **if** $j > i$ **then**
14 | | | $\mathcal{B}_{j-1} = \mathcal{B}_{j-1} \backslash b_j$
15 | | **end**
16 | | $j = j - 1$
17 | **end**
18 **end**
19 **if** $j \geq i$ *and* $|\mathcal{B}_i| > 0$ **then**
20 | Use Algorithm 3 to allocate bands to sections $R_i, \ldots, R_j$, from the free bands sets $\mathcal{B}_i, \ldots, \mathcal{B}_j$
21 **end**
22 Return $i, \mathcal{B}_i$

Proof In Algorithm 4, the first while loop at line 2 breaks out with either (1) $i < n$, or (2) $i = n$, or (3) $i = n + 1$. At this time, sections $R_1, \ldots, R_{i-1}$ have been successfully allocated with bands. Next we analyze these three cases respectively. If the loop breaks out with $i < n$, then we must have $|\mathcal{B}_i| \neq 1$. In the following discussion, when we refer to i, we mean the fixed value of i after the loop breaks out. If $|\mathcal{B}_i| \neq 1$, then it must be true that $|\mathcal{B}_i| > 1$, which can be argued as follows. First of all, if $i = 1$, which may happen only if the first section is a non-SFB section, hence $|\mathcal{B}_i| > 1$. Second, if $1 < i < n$, then the original $\mathcal{B}_i$ has at least two free bands. Line 5 removes only one band from $\mathcal{B}_i$. Hence after the band removal by line 5, we must have $|\mathcal{B}_i| > 1$, since we already have $|\mathcal{B}_i| \neq 1$. If the first loop breaks out with $i < n$, the second while loop (backward scan) will be executed and breaks out with either $j = i - 1$ or $j \geq i$. This backward scan starts at section R_n and goes backwards section R_i. In the following discussion, when we refer to j, we mean the fixed value of j after the second while loop breaks out. If $j = i - 1$, sections $R_i, \ldots, R_n$ have been successfully allocated with bands by the second while loop, and the allocated band of each section is different from the one for the downstream section. On the other hand, sections $R_1, \ldots, R_{i-1}$ have been successfully allocated with bands by the first while loop, and the allocated band of each section is different from the one for the upstream section. Furthermore, in the first while loop, b_{i-1} has already been excluded from $\mathcal{B}_i$. Hence the allocated band of R_i is different from the ones of the upstream and downstream sections R_{i-1} and R_{i+1}, b_{i-1} and b_{i+1}. Thus, there exists a band allocation for the path segment satisfying Definition 5.1. If $j \geq i$, the second while loop must break out due to $|\mathcal{B}_j| > 1$, and has successfully allocated bands to sections $R_{j+1}, \ldots, R_n$. On the other hand, sections $R_1, \ldots, R_{i-1}$ have been successfully allocated with bands by the first while loop. The $\mathcal{B}_{i+1}, \ldots, \mathcal{B}_{j-1}$ are not modified and hence have at least two free bands since they are internal sections of the path segment. Based on the earlier discussions, $\mathcal{B}_i$ and $\mathcal{B}_j$ also have at least two bands. Hence, Algorithm 3 can successfully allocate bands for sections $R_i, \ldots, R_j$. Furthermore, the allocated band of R_i must be different from b_{i-1}, since b_{i-1} has been excluded from $\mathcal{B}_i$ in the first while loop. Similarly the allocated band of R_j is also different from b_{j+1}, since it has been excluded from $\mathcal{B}_j$ in the second while loop. Therefore, each section has an allocated band different from the upstream and downstream sections, i.e., there exists a band allocation for the path segment. In conclusion, if the first while loop breaks out with $i < n$, there exists a band allocation for the path segment. With the second case, i.e., if the first while loop breaks out with $i = n$, then it breaks out due to either $|\mathcal{B}_n| = 0$ or $|\mathcal{B}_n| > 1$. This is because if $|\mathcal{B}_n| = 1$, the loop would continue to break out with $i = n + 1$. Now if $|\mathcal{B}_n| > 1$, then line 19 will use Algorithm 3 to allocate a band for section R_n to complete the band allocation. Hence, there exists a band allocation for the path segment if the first while loop breaks out with $i = n$, and $|\mathcal{B}_n| > 1$. If $|\mathcal{B}_n| = 0$, then there is no free band to be allocated to section R_n. Furthermore, when the first while loop allocates band to each section among $R_1, \ldots, R_{n-1}$, there was only one choice as the updated $\mathcal{B}_i$ has only one band. Hence, $b_1, \ldots, b_{n-1}$ are the only band allocation for sections $R_1, \ldots, R_{n-1}$ to satisfy Definition 5.1. As a result, there does not exist a band allocation for the path segment if the first while loop breaks out with $i = n$ and $|\mathcal{B}_n| = 0$. Furthermore, this is also the only case that there is no band allocation. With the third case, i.e., if the while loop breaks out with $i = n + 1$, then the first while loop has already successfully allocated a band to each section of the path segment.

By Theorem 5.1, Algorithm 4 can be used to independently allocate band for each path segment. Algorithm 4 scans each section only once. Since collecting the free bands for each section just needs to scan each node once, then the time complexity of Algorithm 4 is $\Theta(N)$, where N is the number of nodes of the path segment. We call an algorithm *optimal* if it can find the band allocation for the sections on a path whenever there exists one, and the time complexity is $\Theta(N)$. Thus, the band allocation algorithm for the PP-SB-IM service is optimal.

5.4.1 Analysis of Band Allocation for PP-SB-IM Service

Next, we analyze the blocking probability of a path segment for a PP-SB-IM demand. We first introduce the necessary and sufficient condition for a successful band allocation for a path segment in the following theorem.

Theorem 5.4 *There is no band allocation for a path segment $\{R_1, \ldots, R_n\}$ iff (a) R_1 and R_n are SFB sections, and all internal (non-SFB) sections have exactly two* *free bands; and (b) the band allocation algorithm is able to successfully allocate bands to sections $R_1, \ldots, R_{n-1}$, where one of the free bands of R_i ($2 \le i \le n-1$) is the allocated band of R_{i-1}, but fails to allocate a band to R_n, because its only free band is the allocated band of section R_{n-1}.*

Proof By Theorem 5.3, the band allocation for a path segment fails if and only if Algorithm 4 terminates with $i = n$ and $|\mathcal{B}_n| = 0$. From the proof of Theorem 5.3, in this case ($i = n$ and $|\mathcal{B}_n| = 0$), sections $R_1, \ldots, R_{n-1}$ have been successfully allocated with bands by the first while loop of Algorithm 4. Furthermore there is one and only one band allocation for sections $R_1, \ldots, R_{n-1}$, which can happen only if R_1 is an SFB section, sections $R_2, \ldots, R_{n-1}$ each has exactly two free bands, where one of the free band is the allocated band of the upstream section. Algorithm 4 fails to allocate a band for section R_n because after removing the allocated band of R_{n-1} from $\mathcal{B}_n$, $\mathcal{B}_n$ becomes empty ($|\mathcal{B}_n| = 0$) and the algorithm exits. Note that originally (before updating $\mathcal{B}_n$) we must have $|\mathcal{B}_n| = 1$. This implies that the band in the original $\mathcal{B}_n$ is the allocated band of R_{n-1}.

For a path segment with n sections $R_1, \ldots, R_n$, when sections $R_2, \ldots, R_{n-1}$ each has exactly two free bands, by Theorem 5.4, the band allocation fails if for each section among $R_2, \ldots, R_{n-1}$, one of the two free bands is the allocated band of the upstream section, and the only free band of R_n is the allocated band of R_{n-1}. By the random band allocation of Algorithm 4, the two free bands of section R_u ($2 \le u \le n-1$) should be randomly distributed among the M bands. Then the probability that one of the two free bands of section R_u is the allocated band of section R_{u-1} is equivalent to the probability that when distributing two balls into M boxes, one of the balls falls into a given box (the allocated band of R_{u-1}). This probability is $2 \times \frac{(M-1)}{M(M-1)} = \frac{2}{M}$. Moreover, the probability that the only free band of section R_n is the allocated band of section R_{n-1} is $\frac{1}{M}$. We assume that the free band distribution is independent between sections. Let y_u denote the number of free bands of section R_u ($1 \le u \le n$). Let $\mathbf{y}$ denote the vector $[y_u]_{1 \le u \le n}$. Let $q(\mathbf{y})$ denote the conditional blocking probability of the path segment for PP-SB-IM service given $\mathbf{y}$. From the above discussions, we can write $q(\mathbf{y})$ as

$$q(\mathbf{y}) = \begin{cases} \frac{1}{M} \left(\frac{2}{M}\right)^{n-2}, & \text{if } n \ge 2, y_1 = y_n = 1, \text{ and } y_u = 2 \text{ for } 1 < u < n \\ 0, & \text{otherwise} \end{cases} \qquad (5.5)$$

Next we assume that a path p has n sections, also denoted as $R_1, \ldots, R_n$. Let l_u denote the number of free bands of section R_u. Let $\mathbf{l}$ denote the vector $[l_u]_{1 \le u \le n}$. We can group the path sections into path segments. Let $\mathcal{K} = \{u \mid l_u = 1, 1 \le u \le n\}$ denote the SFB sections on path p. Let $u_1, \ldots u_{|\mathcal{K}|}$ denote the section numbers in $\mathcal{K}$, at the ascending order. If $u_1 > 1$, then R_1 is a non-SFB section, and by Theorem 5.2, the blocking probability of the first segment $\{R_1, \ldots, R_{u_1}\}$ is 0. Similarly, if $u_{|\mathcal{K}|} < n$, then R_n is a non-SFB section, and the blocking probability of the last segment $\{R_{u_{|\mathcal{K}|}}, \ldots, R_n\}$ is also 0. Hence we may only need to consider $|\mathcal{K}| - 1$ segments, partitioned by $|\mathcal{K}|$ SFB sections. Let $\mathbf{l}_k$ denote the vector of the number of free bands for the kth segment. Let $\sigma(\mathbf{l}, \mathcal{K})$ denote the conditional blocking probability of path p for a PP-SB-IM demand, given $\mathbf{l}$ and $\mathcal{K}$. Then by Theorem 5.4, $\sigma(\mathbf{l}, \mathcal{K})$ can be obtained as follows.

$$\sigma(\mathbf{l}, \mathcal{K}) = \begin{cases} 1, & \text{if } \exists u, l_u = 0 \\ 0, & \text{if } l_u \geq 2 \text{ for } 1 \leq u \leq n \\ 1 - \prod_{k=1}^{|\mathcal{K}|-1} (1 - q(\mathbf{l}_k)), & \text{otherwise} \end{cases}$$

Let $\theta_u(l_u)$ denote the probability that there are l_u free bands at section R_u. It can be computed by applying (5.3) to the nodes on section R_u, with $\beta(0 \mid \ldots)$ replaced by $\beta(l_u \mid \ldots)$. Then the blocking probability of path $p = \{R_1, \ldots, R_n\}$, denoted as σ^p, is given as

$$\sigma^p = \sum_{l_1=0}^{M} \cdots \sum_{l_n=0}^{M} \prod_{u=1}^{n} \theta_u(l_u) \sigma(\mathbf{l}, \mathcal{K}), \tag{5.6}$$

where $\mathcal{K}$ changes according to $\mathbf{l}$.

Let $\mathcal{R}(v)$ denote the set of path sections that contain node v. Let $\mathcal{P}(v)$ denote all paths that have one or more sections containing node v. Let $\theta_u(l_u \mid v, k)$ denote the conditional probability that there are l_u free bands at section R_u, given that there are k free bands at node v. The $\theta_u(l_u \mid v, k)$ is similarly computed as $\theta_u(l_u)$, by restricting the value of i_v. Let $\sigma_v^p(k)$ denote the conditional blocking probability of path $p \in \mathcal{P}(v)$, given that there are k free bands at node v. $\sigma_v^p(k)$ is computed by replacing $\theta_u(l_u)$ with $\theta_u(l_u \mid v, k)$ in (5.6).

Let $\mathcal{P}_R$ denote the paths that traverse section R. Given that there are k free bands at node $v \in R$, the arrival rate of carried demands at section R, denoted as $\omega_v(R \mid k)$, is given as $\omega_v(R \mid k) = \sum_{p \in \mathcal{P}_R} \lambda_p(1 - \sigma_v^p(k))$. The arrival rate of carried demands at node v given that the number of free bands is k, i.e., $\rho_v(k)$, can be computed as follows.

$$\rho_v(k) = \sum_{R \in \mathcal{R}(v)} \omega_v(R \mid k) \tag{5.7}$$

Finally, we can solve the nonlinear Eqs. (5.6–5.7) and (5.4) to obtain σ^p, and then compute the average blocking probability for PP-SB-IM service, similarly as in last section.

5.5 Performance Evaluation

We evaluate the performance of the ODSA architecture through both the analytical model and simulations. The sample network is a 16-node Grid network. We focus on 10 node pairs $(1, 4)$, $(1, 13)$, $(1, 16)$, $(2, 14)$, $(3, 15)$, $(4, 13)$, $(4, 16)$, $(5, 12)$, $(8, 9)$, and $(13, 16)$, where nodes are numbered in row major order in the Grid. The routing path between each node pair is a randomly picked shortest path. (There are multiple shortest paths between a node pair.) The number of bands M is assumed 20. The user demands are assumed to follow the Poisson arrival, and have exponentially distributed

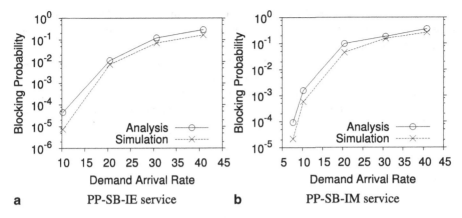

Fig. 5.4 Blocking probability for PP-SB-IE and PP-SB-IM services, with random demands

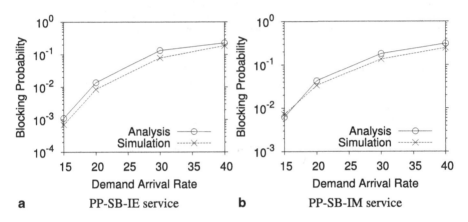

Fig. 5.5 Blocking probability for PP-SB-IE and PP-SB-IM services, with uniform demands

durations. Without loss of generality, let the mean duration of user demands be one time unit. The arrival rate of user demands for each node pair (s, d) is generated as $\lambda_{sd} = \tau \chi_{sd}$, where τ is a scale parameter to control the total demand amount. χ_{sd} is a random number in $(0, 1)$ or a constant 1. In the former case $(\chi_{sd} \in (0, 1))$, we call it 'random demand', and in the latter case, we call it 'uniform demand'. The simulation time for obtaining the simulation results is $10, 000$ s.

Figure 5.4 plots the blocking probability for the PP-SB-IE and PP-SB-IM services, obtained by analysis and simulation, respectively, with the random demand. We can see that the analysis results match simulation results well, although there is some overestimation due to the independence assumption for the number of free bands at different nodes. From the figure, we can also estimate the network capacity. For the PP-SB-IE service, when the demand arrival rate is 20–30, the blocking probability is around 0.01–0.1. This means the network can accommodate about 20–30 demands per time unit very well. The blocking probability for the PP-SB-IM service is higher than for the PP-SB-IE service. This is because in the former case, each switching node needs to reserve two bands.

Figure. 5.5 illustrates the blocking probability with the uniform demand. Comparing Figs. 5.4 and 5.5, one can see that the blocking probability primarily depends on the demand arrival rate. The variation of arrival rates among different node pairs in the case of the random demand does not have a significant effect on the blocking probability.

References

1. Buddhikot M, Ryan K (2005) Spectrum management in coordinated dynamic spectrum access based cellular networks. Proc. IEEE DySPAN, pp 299–307
2. Subramanian A, Gupta H, Das SR, Buddhikot MM (2007) Fast spectrum allocation in coordinated dynamic spectrum access based cellular networks. Proc. IEEE DySPAN, pp 320–330
3. Redi J, Ramanathan R (2011) The DARPA WNaN network architecture. Proc. MILCOM, pp 2258–2263
4. Nandagopal T, Kim T-E, Gao X, Bharghavan V (2000) Achieving mac layer fairness in wireless packet networks. Proceedings of ACM MobiCom, pp 87–98
5. Ramanathan S, Lloyd E (1993) Scheduling algorithms for multihop radio networks. IEEE/ACM Trans Netw 1(2):166–177
6. Barrett C, Istrate G, Kumar V, Marathe M, Thite S, Thulasidasan S (2006) Strong edge coloring for channel assignment in wireless radio networks. IEEE Pervasive Computing and Communications (PerCom) Workshops
7. Krumke SO, Marathe MV, Ravi SS (2000) Models and approximation algorithms for channel assignment in radio networks. Wirel Netw 7:575–584

Chapter 6
Conclusions

Thanks to the exponential growth of wireless devices and wireless traffic, the demand for radio spectrum is stronger than ever. However, the radio spectrum has been already largely licensed, due to the static spectrum allocation policy. This creates the so-called spectrum scarcity problem. Fortunately, the spectrum scarcity is an artificial problem. Many studies have found that the licensed spectrum is significantly under-utilized in the temporal, spatial, and frequency domains. Motivated by these findings, there have been extensive research on how to significantly increase spectrum utilization in recent years. A consensus from the research community is to allow users to share spectrum dynamically and efficiently.

In this book, we have introduced several dynamic spectrum sharing architectures. The first architecture, called the *opportunistic spectrum access* (OSA), has been studied and adopted in most studies in the last decade. In this architecture, secondary users (SUs) use cognitive radios to sense the spectrum environment to search for idle spectrum bands that are not being used by primary users (PUs), and opportunistically access those bands. On the other hand, when a PU signal re-appears on a spectrum band, the SUs that are currently accessing the band must vacate immediately, to avoid harmful interference to the PU. The OSA architecture provides an effective way for the SUs to dynamically access the unused spectrum and thus can potentially increase spectrum utilization significantly. However, the re-appearance of PU signals arbitrarily disrupts the SU communications. This results in high overhead for SUs and the performance for SUs can be highly unstable.

To address the issue of the OSA architecture, several new spectrum sharing architectures have been proposed recently. In Chaps. 3 and 4, we present two architectures, IC-DSA and DSCA, which allow SUs to simultaneously access spectrum with PUs, through providing incentives to PUs. The IC-DSA architecture exploits the network coding technology to 'piggyback' SU packets onto PU packets. In the DSCA architecture, an SU splits a portion of its power to help to boost the *signal to interference plus noise ratio* (SINR) of the PUs, to offer incentives, and utilizes the dirty paper coding technology to reduce interference. In the IC-DSA and the DSCA architectures, the performance of both the PU and the SU is increased, creating a win-win situation. The difference between the two architectures is that in the DSCA architecture, the SU transmissions are transparent to the PUs, while in the IC-DSA architecture, the

© The Author(s) 2015
C. Xin, M. Song, *Spectrum Sharing for Wireless Communications,*
SpringerBriefs in Electrical and Computer Engineering, DOI 10.1007/978-3-319-13803-9_6

PUs need to decode the packets from SUs to utilize network coding between both. At last, we have introduced an application-oriented spectrum sharing architecture called *on-demand spectrum access* (ODSA). This architecture enables users to set up application-oriented virtual topologies to support specific user applications, such as a video conference, or transfer of a large data file. It dynamically and efficiently allocates spectrum to users through an optimal band allocation algorithm, so that the spectrum is efficiently utilized among all users while the technical complexities such as spectrum sensing can be avoided.

A recent development in spectrum sharing is the *licensed shared access* [1]. This architecture is primarily designed to appeal to mobile operators. The SUs, i.e., the mobile operators in this case, need to be licensed to share spectrum bands with the incumbent users (PUs). The mobile operators may still need to pay a certain amount of fee to obtain the license as in the spectrum auction. Nevertheless, this fee is expected lower than the exclusive license. The spectrum sharing is similar to the case of TV white space. Typically, a license repository, like a spectrum database, stores the information of available spectrum bands in multi-dimensions. A license controller is responsible to assign these bands to the licensed SUs (mobile operators). The licensed shared access architecture seeks to not only provide protections to PUs from interference, but also offer some guarantee for the *quality of service* (QoS) for the mobile operators. The latter requirement implies that the spectrum usage by the incumbent users should be somehow predictable, so that the spectrum access by the mobile operators is predictable, to offer predictable QoS.

Due to its potential to significantly increase spectrum utilization to meet the ever-increasing spectrum demand, dynamic spectrum sharing has attracted extensive research, and will continue to do so in the future, to meet the fast growth of wireless devices, wireless traffic, and wireless applications. The breakthrough in this field will have a fundamental and profound impact on the future wireless communications systems and technologies.

Reference

1. Matinmikko M, Okkonen H, Palola M, Yrjola S, Ahokangas P, Mustonen M (2014) Spectrum sharing using licensed shared access: the concept and its workflow for lte-advanced networks. IEEE Wirel Commun 21(2):72–79